EVERYMAN, I will go with thee,
and be thy guide,
In thy most need to go by thy side

Mathematics for Everyman

Laurie Buxton

Diagrams by Claire Buxton

J. M. Dent & Sons Ltd
London Melbourne

First published 1984
First paperback edition 1985

Printed in Great Britain by
The Guernsey Press Co. Ltd, Guernsey, C.I., for
J. M. Dent & Sons Ltd
Aldine House, 33 Welbeck Street, London W1M 8LX

British Library Cataloguing in Publication Data

Buxton, Laurie
Mathematics for everyman.—(Everyman's reference library)
1. Mathematics—1961-
510 QA37.2

ISBN 0-460-02402-7

For 'Everywoman' as well as 'Everyman'

Acknowledgments

To my wife, Pam, for typing from my illegible hand, and correcting my sums.
To my daughter Claire, for doing the diagrams.
To the people who wrote the books I have read.

Contents

Introduction

Mathematics is a living and growing organism; within it are intricate and delicate structures of strong aesthetic appeal. It offers opportunities for surprise as unexpected vistas open the mind to new lines of thought. There is a moment when, after struggling with an intractable problem, the mechanism is suddenly clear, and the problem resolves with a deep sense of rightness. In this it offers the most intense of intellectual pleasures.

Mathematics was created by all manner of people. There were religious bigots and atheists, political reactionaries and wild revolutionaries, snobs and egalitarians; some were people of great charm, some odious. If there is any common denominator, it is a driving curiosity, a desire to understand, a need to build, even if the structures be abstract. Admirably suited though mathematics is to modelling the real world, it can be developed totally without dependence on anything outside itself. Parts of it are simply mind creations, owing nothing to the physical world. It is a playground for the mind.

Throughout the centuries it has always been recognized as one of the central strands of human intellectual activity. Never has its importance been questioned, though few could properly characterize its specific significance. As we shall see, it may, with language, be the activity most closely related to the way we think.

By no means all of mathematics is intellectually stimulating. It is the underlying ideas that give it its special quality, yet vast tracts of meaningless manipulation have been presented as if they were of importance in the subject. Increasingly this work is being taken over by machines, and this very fact separates the wheat from the chaff. If something can be done by a machine, let it be. What is left is what interests the active mind. We are at the beginning of great changes resulting from the impact, first of the pocket calculator and then of the microcomputer. Where it will lead we can only speculate.

Regettably, many of us have never been allowed to see what mathematics is. It has been obscured by pointless emphasis on routines rather than ideas. This failure to distinguish what is important has led many people to see mathematics as a collection of totally arbitrary rules which have to be learnt by rote, and performed with the exactness and precision of a religious rite. Ask a person if there is much to be remembered in mathematics; if they speak of an overwhelming mass of material, their education in this area has been counter-productive, not merely neutral. Mathematics, properly seen, is a connected web; grasp at one piece and all the surrounding region comes to mind.

The general attitude of the population to mathematics is very negative. In a recent survey people were approached at random to discover what mathematics they used in everyday life. About half flatly refused to have anything to do with someone who proposed to discuss mathematics with them. Assurances that no sort of testing was involved were of no avail.

In this book we hope to resolve some of these problems. The intent is to give a broad-ranging view of what the subject is about, constantly returning to known territory, showing the historical trends, and opening a wider and more flexible view of each mathematical topic.

There are four main sections to the book. In the four chapters which form the first section we seek to throw new light on what people call the 'basics', then widen the discussion to what we might call 'citizen's' mathematics, tell the story of calculation down the ages, and then have a chat with a microcomputer, to which we shall turn from time to time throughout the other chapters.

Our second section starts with the traditional strands of mathematics as we remember them from our schooldays, number, geometry and algebra. With each we start back at the beginnings, in very familiar territory, but try to show how novel ideas constantly infuse even the most ancient of studies with new life.

We move from counting to the mysteries of the square root of minus one, from Euclid to the space-time of Einstein, and from simple equations to Boolean algebra. If we can gain an overview of these subjects, the filling in of more detailed work is eased. An overall framework and purpose allows us to tackle minutiae without wondering what it is all for.

Next in this section we tackle some less familiar topics, considered advanced or modern. In our chapter on the calculus we try to de-mystify the notion of infinitesimals through the use of modern calculating devices. We show it as a working tool based on firm common sense. We then turn to probability theory, at one end the basis of gambling and at the other a deeply philosophical issue about the way the world and we ourselves are. Like calculus, it does not date from ancient times, yet it would not be called modern. The final chapter in this section takes two modern topics from the many available, sets and matrices, and puts them in context both in the real world and within mathematics itself.

Our third section examines mathematics in two of its important facets. The first of these two chapters sees mathematics interacting with the modern technological world and providing new ways of solving a range of problems. We introduce operational research and graph theory and show how they develop in response to needs in society. The second chapter looks more inwardly and tries to give the reader some feel for what it is like to explore genuinely within mathematics. Here the relationship is between mathematics and the mind rather than the outside world.

Finally we try to say something of what mathematics is. We attempt to place it philosophically in relation to other disciplines, in relation to language, and to the way our minds work.

Throughout there are opportunities for those who wish to push further, to tackle problems and to look at more mathematically demanding issues. Such matters have been collected in an appendix, in the hope that the main text may be read with some fluency, while offering more entertainment for 'afters'.

The intent has been to avoid the extensive use of symbols. This is not easy in a subject so dependent on them, but we have sought to explain in prose as opposed to equations and formulae. If readers can gain some feeling of confidence that they know what the whole scene is about, then perhaps they may return to mathematics. If, however, by the end of the book they at least feel that the subject is more interesting than they had thought, then an aim has been achieved.

I

Starting out

1

Back to basics

There is something satisfying in the slogan 'Back to Basics'. The word 'basic' has a comforting ring to it. It is also a word often used by a person trying to win over another to his particular point of view. There is a lot to be said for learning the basics, but it helps to know what they are before you start.

There is a generally received view in the population as to what constitutes the basics in mathematics. We have the multiplication tables and the four rules – and do we perhaps extend them to fractions? This view would not in general be shared by those who know much about mathematics. Their significance, though, lies in the fact that so many do believe in their centrality, rather than whether that belief is true.

Is there anything new we might say about the multiplication tables? Though it may seem unlikely it is worth a try. At one time it was the practice to festoon the classroom walls with them, though it was necessary to have a way of covering them for the mental arithmetic test. The first obvious problem if the tables are presented in this way is the sheer volume. There are so many separate facts to know and to learn by rote. Though human young have a good deal of room for input, overload is possible, and many suffer from it. As we moved towards metrication it became more the practice only to go as far as 10×10. It is interesting to see what proportion of the tables this knocks out (see p. 3). The quantity seems so excessive in part because the tables are here presented as separate, unconnected facts. Chanting them is not very helpful, since $7 \times 9 = 63$ sounds no more poetic than $7 \times 9 = 57$. In fact learning by rote without any attempt at organizing the facts is not sensible. We can and should distinguish between memorizing and learning by rote, and there is no doubt that some material must be memorized. Satisfactory memorizing implies working over the material until it is in a connected and related form, before we lodge it in our minds.

We can make a start by revising our multiplication facts into the form shown at the top of p. 4. To find a product we need to read off on the two scales. To find 5×6 we look for 5 on the base and 6 up and there is no number at this intersection. However 6×5 is there, recorded as 30.

We have greatly reduced the volume, but introduced a good deal more structure. We have first dispensed with half the numbers, on the grounds that if we know 9×7 we do not need to fill in 7×9. This gives us 45 numbers in the

1×1 =1	2×1 =2	3×1 =3	4×1 =4	5×1 =5	6×1 =6
1×2 =2	2×2 =4	3×2 =6	4×2 =8	5×2 =10	6×2 =12
1×3 =3	2×3 =6	3×3 =9	4×3 =12	5×3 =15	6×3 =18
1×4 =4	2×4 =8	3×4 =12	4×4 =16	5×4 =20	6×4 =24
1×5 =5	2×5 =10	3×5 =15	4×5 =20	5×5 =25	6×5 =30
1×6 =6	2×6 =12	3×6 =18	4×6 =24	5×6 =30	6×6 =36
1×7 =7	2×7 =14	3×7 =21	4×7 =28	5×7 =35	6×7 =42
1×8 =8	2×8 =16	3×8 =24	4×8 =32	5×8 =40	6×8 =48
1×9 =9	2×9 =18	3×9 =27	4×9 =36	5×9 =45	6×9 =54
1×10=10	2×10=20	3×10=30	4×10=40	5×10=50	6×10=60
1×11=11	2×11=22	3×11=33	4×11=44	5×11=55	6×11=66
1×12=12	2×12=24	3×12=36	4×12=48	5×12=60	6×12=72
7×1 =7	8×1 =8	9×1 =9	10×1 =10	11×1 =11	12×1 =12
7×2 =14	8×2 =16	9×2 =18	10×2 =20	11×2 =22	12×2 =24
7×3 =21	8×3 =24	9×3 =27	10=3 =30	11×3 =33	12×3 =36
7×4 =28	8×4 =32	9×4 =36	10×4 =40	11×4 =44	12×4 =48
7×5 =35	8×5 =40	9×5 =45	10×5 =50	11×5 =55	12×5 =60
7×6 =42	8×6 =48	9×6 =54	10×6 =60	11×6 =66	12×6 =72
7×7 =49	8×7 =56	9×7 =63	10×7 =70	11×7 =77	12×7 =84
7×8 =56	8×8 =64	9×8 =72	10×8 =80	11×8 =88	12×8 =96
7×9 =63	8×9 =72	9×9 =81	10×9 =90	11×9 =99	12×9 =108
7×10=70	8×10=80	9×10=90	10×10=100	11×10=110	12×10=120
7×11=77	8×11=88	9×11=99	10×11=110	11×11=121	12×11=132
7×12=84	8×12=96	9×12=108	10×12=120	11×12=132	12×12=144

The multiplication tables

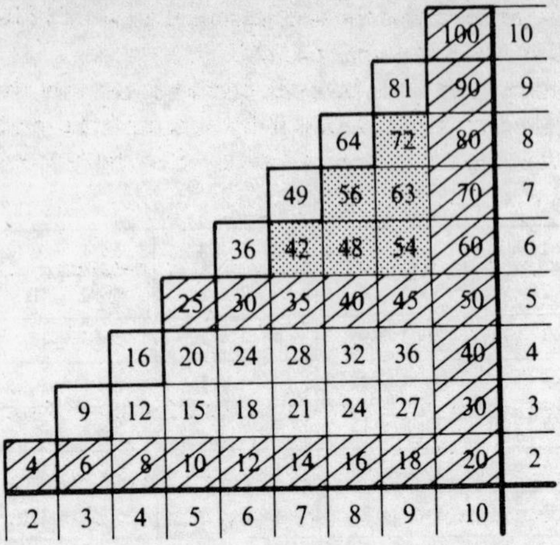

A table triangle

table. That is how many facts we need to know. We have shaded the 2 ×, 5 ×, 10 × tables with oblique strokes, not on the grounds that we need not know them, but that they are reasonably easy and few people have trouble with them. Master those 22 facts and we are down to 23 facts.

Look now at the steps up the slope . . . 4, 9, 16, 25, 36, etc. These are very interesting numbers and we shall take time off from learning our tables to look at them more closely. Each is along the diagonal of our table square and is found by multiplying a number by itself.

$$36 = 6 \times 6$$
$$81 = 9 \times 9$$
$$25 = 5 \times 5$$

The reason we call them squares is obvious enough. If we have 36, 81 or 25 dots we can arrange them in a square:

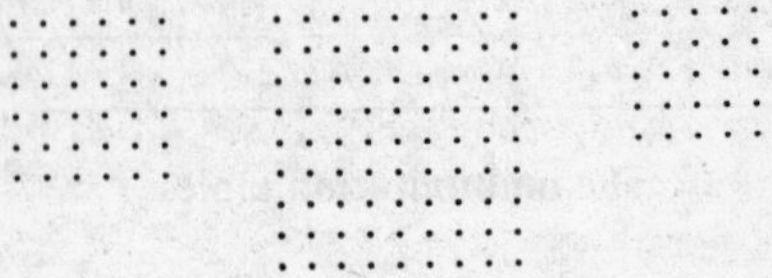

The squares are interesting in another way. If we add up this string of odd numbers, starting with 1, we always get a square. For instance

$$1+3+5+7+9+11+13 = 49$$

and we know that we shall find the next square at a gap of 15 further on at 64. This is discussed in the Appendix (p. 228).

There are many such links between numbers, not just the squares. The 9× table is another example. Learn it by rote and the pattern is missed. Observe the numbers in the column above the 9 in the bottom line. They go

18, 27, 36, 45, 54, 63, 72, 81

In every number the two digits add up to 9. For instance

$$36:\quad 3+6 = 9$$
$$72:\quad 7+2 = 9$$

There is reason behind this, as there was in the addition of the odd numbers to give squares. In the reasons lies the interest. You will find some more explanation of these links in the Appendix (pp. 228–9); the point is that there are patterns everywhere. It is in observing pattern that the number facts become coherent, rather than the separate statements of our tables.

Return now to our table triangle on p. 4. We had reduced the issue to 23 facts. By observing and working with the square numbers we gradually learn them. We learn them not by rote but by observing links and connections, until the facts become natural to us. Knowing about the squares, for instance, prevents us ever giving an answer such as $8 \times 7 = 49$.

We next learn, by practice, by using a number line, and by other means, the 3× and the 4× tables. They are not generally thought specially difficult – yet let us admit that somehow we need to commit them to memory. What is now left simply does not seem the same sort of task as we first presented.

We are left with the dark squares. Six facts, three of them in the 9× table, which has special features. Not so monumental a task when sorted out in this way. The activity of understanding and seeing connections has the effect of processing the facts to make them digestible. Then we memorize them as a neat package that takes up a lot less room in our minds.

Giant's Causeway

The photograph on p. 6 is of a wood sculpture by Brian Willsher. The vertical rods are made of white pine, approximately one inch square. The upward sweeping effect is dramatic and the shadows as light is thrown from different angles are intriguing. It has attracted keen attention from many with an aesthetic response to shape and form. The sculpture is a three-dimensional model of the multiplication tables.

Look on p. 7 at our complete table square. It is the extension of the triangle used earlier. This time 30 appears opposite 6 in the 5 column and

The Giant's Causeway: 3-dimensional multiplication tables

opposite 5 in the 6 column. It is a standard two-way table used in many schools and called a 'table square'.

10	20	30	40	50	60	70	80	90	100
9	18	27	36	45	54	63	72	81	90
8	16	24	32	40	48	56	64	72	80
7	14	21	28	35	42	49	56	63	70
6	12	18	24	30	36	42	48	54	60
5	10	15	20	25	30	35	40	45	50
4	8	12	16	20	24	28	32	36	40
3	6	9	12	15	18	21	24	27	30
2	4	6	8	10	12	14	16	18	20
1	2	3	4	5	6	7	8	9	10

The table square

We used only half in our previous diagram because the numbers on each side of the diagonal are symmetric. It is this symmetry that has artistic appeal in our Giant's Causeway, for the main idea in the design of this model is that the pine rods should have heights as specified in the table square on which they stand. The rod standing on the number 45 has height 45 units, and so on.

We have marked the tops of the rods that form the line of symmetry. The heights are, of course, the square numbers, and the central line of the Giant's Causeway sweeps down in a graceful parabola. Pick out all the rods of the same height (say 24) in our table square and we have the beginnings of a hyperbola. These curves are two of the 'conic sections' which the Greeks studied and which we examine further in our chapter 'Mathematics in Action'. The linkages in mathematics between various topics are a source of constant interest and surprise.

Factors, not products

We usually learn the multiplication facts, in one way or another, by coming to know the result of multiplying two numbers together to get their product. There is another way. We can learn to break numbers up into their component parts. If we start with 63 we might say

$$63 = 9 \times 7$$

and then we find that the 9 can also be split, so

$$63 = 3 \times 3 \times 7$$

and now we are down to the hard building-bricks of number. Yet we might have started another way,

$$63 = 3 \times 21$$

and now we can break down the 21,

$$63 = 3 \times 3 \times 7$$

and despite the different route we arrive at the same final result.

That might seem very unsurprising. Through experience it may have become a commonplace that this happens. We may recognize that 36 can break down (as we see in our table square) in a number of different ways, 2×18, 3×12, 4×9, 6×6 . . . yet when we continue the breaking down we always get

$$36 = 2 \times 2 \times 3 \times 3$$

We can rearrange, but there are always two 2s and two 3s. Obvious? It is in fact a very difficult matter to prove. The fact that any number breaks down into the product of *prime* factors in one and only one way is known as the fundamental theorem of arithmetic and has exercised many powerful minds, including that of Gauss, perhaps the greatest mathematician of all time.

It is a sobering thought that many of us spent a vast number of hours on learning the tables when children, but are totally unaware that there is even such a thing as the fundamental theorem of arithmetic.

We all vary in what we find easy and what difficult. Our table facts illustrate two modes of thinking. In the first we build up numbers by combining their parts (the factors). This is synthesizing and is the way most of us learnt the facts. The second is to break down numbers into their components. This process is analysis. It may seem a small matter whether we say $5 \times 7 = 35$ or $35 = 5 \times 7$, yet there is an implication of a very distinct difference in thinking.

We seem to be helped by putting numbers in an organized fashion. The table square has proved to have a number of interesting features. There is another square, even simpler, which can show us some more features of our whole numbers. We call it a hundred square, for obvious reasons. It simply sets out all the numbers from 1 to 100 in rows of ten (see p. 9).

If we tackle multiplication by looking for the answers to products from our tables we hit a number of the little squares. Some squares get hit a lot (like 36). But a surprising number do not get hit by our product approach. They fall into two categories. The first contains numbers such as

57, 58, 87, 91

1	2	3	4	5	6	7	8	9	10
11	12	13	14	15	16	17	18	19	20
21	22	23	24	25	26	27	28	29	30
31	32	33	34	35	36	37	38	39	40
41	42	43	44	45	46	47	48	49	50
51	52	53	54	55	56	57	58	59	60
61	62	63	64	65	66	67	68	69	70
71	72	73	74	75	76	77	78	79	80
81	82	83	84	85	86	87	88	89	90
91	92	93	94	95	96	97	98	99	100

The hundred square

You will not find these as answers in any of our tables but the other approach, factorizing, can give us some insights. We get

$57 = 3 \times 19$ (familiar to darts players)
$58 = 2 \times 29$
$87 = 3 \times 29$
$91 = 7 \times 13$

and we now know something more about them which will not come from the tables. The second category contains this sort of number:

11, 37, 83, 97

and these do not break down. They are called *prime* numbers and play an important role in our thinking about number.

Can we conclude something about the tables? The answer to 'should we learn our tables?' has become clearer. We should know, and have ready to command, all those facts in the tables. It means they must be memorized, but only once the connections have been seen, and not just by rote. We must organize them before we learn them. But we are asking much more. We need to be able to break down all those numbers up to 100 which can be broken down, and to know which ones cannot be broken.

This is a more extensive aim than learning just the tables; and it makes more sense.

The primes

We have just seen that multiplying numbers to find their product is not, in the nature of things, going to interest us in primes. Our reverse approach—to find how to break numbers down – brings them firmly to mind, not just as those which do not break, but as the basic building-bricks of all numbers.

Using the first approach, we are apt to come to regard primes as a sort of 'awkward squad' who do not fit in. The other leads to a much more mathematical view. It sees the primes as the basic numbers. They, not the others, have the central relevance. From them can be formed all the other numbers. They are the building-bricks, or the seeds from which all numbers grow.

In this view 36, 54 and 96 are numbers belonging to the same family. They are all produced from 2s and 3s.

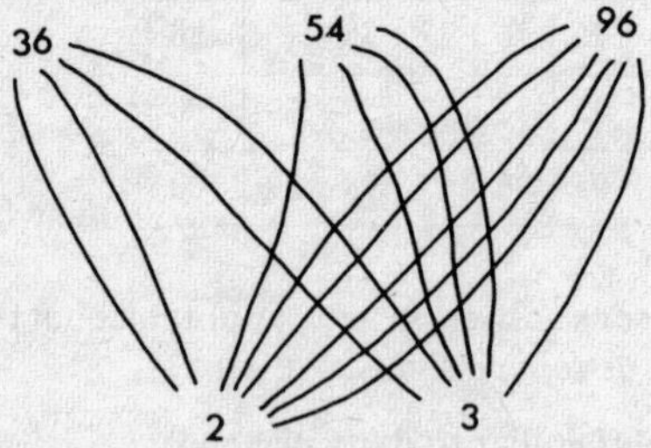

Number families

A game may help us to grasp the way in which larger numbers 'stand on' smaller ones. We start with an array of the first twenty numbers

11	12	13	14	15	16	17	18	19	20
1	2	3	4	5	6	7	8	9	10

(It is easiest to play on a computer, but pencil and paper will do.) The aim is to take numbers totalling as much as possible from the array, and there are only two rules:

(1) The opponent wipes out all the numbers that go into the one which you take.
(2) You must not take a number into which no remaining numbers divide.

You win if you total more than half the total that was originally there. In this case, the numbers 1–20 add up to 210. If we can snatch over 105 we have won.

We start by taking 19; this we can only do because there is a 1 on the array. The 1 now disappears, which means we cannot take another prime – because this would leave the opponent nothing to wipe out. Take in turn the following numbers, crossing out factors as you go:

$$19+4+16+18+15+14+20 = 106$$

Check our moves were legal. We just won – but what is the maximum we could score?

If we increase the array to 60 numbers we have an interesting and demanding game where the image of numbers standing on their prime factors is very helpful. Extra discussion of this game appears in the Appendix (p. 229).

Prime numbers have always fascinated mathematicians, not just because they are so important as the 'atomic structure' of composite numbers, or to use another analogy, the chemical elements as opposed to chemical compounds, but because they raise many questions. When we deal with problems in a later chapter, primes will figure prominently; there are many unsolved problems concerning them. For now, let us find a few basic facts about them.

A prime is a counting number which is divisible by no other counting number. (We exclude one and the number itself.) We commented that

11, 37, 83, 97

were the sort of numbers that did not break down; but we did not prove it. So our first question is 'How do we know if a number is prime?' We need an algorithm (a routine process) for finding out.

Take 11 for example. The brutal, straightforward way is to try to divide it by each of the numbers 2 to 10 in turn. We find that none of them does go exactly and so we know 11 is prime.

Suppose we had 83. We can do the same thing, testing each number from 2 to 82. It would be rather tedious and in doing it we would begin to see that we were wasting a great deal of time. We can, of course, resort to the microcomputer at our elbow, which does not mind wasting time. A very simple program will instruct it to perform this search; we shall look at this later.

Two things occur to us as we work at our divisions. The first is that if we have tested divisibility by 2 and it does not work (in other words we have an odd number) then there is no point in trying to divide by any even number. If 3 does not divide, nor will any multiple of three – we can dispense with the 3 × table taken as high as you like. We arrive at the conclusion that when we are testing for primes, we only need to try primes. This is because of the fundamental theorem. If the number we are looking at does break down, then we can get to its prime factors – and there is one definite string of these. It is only those we need look for.

So on testing 11 we need only have tried 2, 3, 5 and 7. The other numbers below 11 are composite. The task is less tedious, but we still need a list of all these primes. We can certainly build one up; and a computer will help a great deal. But there is another big step in simplifying our process. Go back to 83 and start testing:

83 divided by 2—not exact
83 divided by 3—not exact
83 divided by 5—not exact
83 divided by 7—not exact

The next step would be to try 11. We shall find it does not go, but there is a subtle reason why we did not even need to try it. If we divide 83 by 11 it goes less than 8 times. So if 11 had been a factor, the *other* factor would be less than 8. We know it is not, because we have tried them. The four numbers we have tried determine the fact that 83 is prime.

Look at it a slightly different way. Take one of our square numbers, say 49. Its 'square root' is 7. When we build up 49 we say 7 'squared' is 49 and when we break it down that 7 is the 'square root' of 49. If we divided 49 by a number bigger than 7 the answer (not an exact number) is less than 7. It is at the square root that the two factors 'balance' and become equal.

If we have a number such as 843 to test, we find (on a calculator) that its square root is just over 29. So we only need to test the prime numbers to 29. In fact we are in luck because it divides by 3. The other factor is 241 and if we wish to test whether that is prime we only need go up to 13. A modern computer can test a 100-digit number in 40 seconds.

With enormous numbers, even this relatively quick method becomes very difficult, but it very rapidly provides us with, say, all the primes up to 100.

Eratosthenes devised what he called a 'sieve' for sorting out the primes from the composites. Our 100 square looks exactly like a sieve, so shake it about to allow all the multiples of 2 to fall through. It now looks like the first diagram of our table patterns. The pattern produced by letting all the even numbers (except 2 itself) fall through is not exciting. If we pull the lever to let through the multiples of 3, 5 or 7 we get our next three diagrams. The last of these at least begins to show a pattern with some interest. The patterns help us with our multiplication facts, but also demonstrate once more how deeply rooted is pattern in mathematics.

Set out how ever many numbers you like in rows of equal length (but however long you like) and then march through in equal steps as in a multiplication table. A discernible pattern will always appear, sometimes dull like the 2s and 5s, sometimes more intriguing, more difficult to see. Pattern cannot be avoided.

Now in our 100 square open the flood gates and let out all the multiples of 2, 3, 5 and 7. We are left with those in our last diagram. The sieve of Eratosthenes has produced for us the primes below 100. Again we did not need to go beyond 7, for 11, the next prime, is above the square root of 100.

We have seen pattern in our number multiples and it is natural to look for a pattern in our last diagram. It is, after all, the result of removing a number of overlapping patterns. Perhaps it is an issue more for the artist than the mathematician whether what is left after removing patterns is a pattern. The mathematician has discovered no pattern in the primes. There is not only no formula that produces the primes, there is not even one that produces only primes (however many it leaves out).

So in a central feature of the most basic level of mathematics there is no pattern, no form, nothing we can predict; and that is worrying. That is not to

1	2	3		5		7		9	
11		13		15		17		19	
21		23		25		27		29	
31		33		35		37		39	
41		43		45		47		49	
51		53		55		57		59	
61		63		65		67		69	
71		73		75		77		79	
81		83		85		87		89	
91		93		95		97		99	

1	2	3	4	5		7	8		10
11		13	14		16	17		19	20
	22	23		25	26		28	29	
31	32		34	35		37	38		40
41		43	44		46	47		49	50
	52	53		55	56		58	59	
61	62		64	65		67	68		70
71		73	74		76	77		79	80
	82	83		85	86		88	89	
91	92		94	95		97	98		100

1	2	3	4	5	6	7	8	9	
11	12	13	14		16	17	18	19	
21	22	23	24		26	27	28	29	
31	32	33	34		36	37	38	39	
41	42	43	44		46	47	48	49	
51	52	53	54		56	57	58	59	
61	62	63	64		66	67	68	69	
71	72	73	74		76	77	78	79	
81	82	83	84		86	87	88	89	
91	92	93	94		96	97	98	99	

1	2	3	4	5	6	7	8	9	10
11	12	13		15	16	17	18	19	20
	22	23	24	25	26	27		29	30
31	32	33	34		36	37	38	39	40
41		43	44	45	46	47	48		50
51	52	53	54	55		57	58	59	60
61	62		64	65	66	67	68	69	
71	72	73	74	75	76		78	79	80
81	82	83		85	86	87	88	89	90
	92	93	94	95	96	97		99	100

1	2	3		5		7			
11		13				17		19	
		23						29	
31						37			
41		43				47			
		53						59	
61						67			
71		73						79	
		83						89	
						97			

The sieve of Eratosthenes

say we know nothing about the primes; some very interesting facts have been established.

A fact that we all became aware of early on in life is that the numbers go on for ever; we may remember the feeling of wonder mixed with some disquiet when we first realized this. Do the primes go on for ever? As we go up the numbers, the primes seem to get a bit thinner on the ground, but that is not to say they ever run out, even though we have more and more lower numbers that might go into the number we are testing.

We normally think of Euclid in terms of geometry, but in his *Elements* he gave a simple and elegant proof that they do in fact go on for ever. Yet even nowadays, if anyone were to discover this proof for themselves, it would be evidence of an extraordinary mathematical talent.

Suppose we believed that the primes stopped at 7 (just suppose). Then the number made up by multiplying all the primes and adding 1 is

$$2 \times 3 \times 5 \times 7 + 1$$

Now this number does not divide by any prime, because each of them leaves remainder 1. So *if* 2, 3, 5 and 7 were all the primes, then we arrive at a contradiction, since this new number is a prime. The logic of this and in particular the use of the conditional is quite sophisticated, and still can cause us some thought. Notice that we do not state that $2 \times 3 \times 5 \times 7 + 1$ is a prime, we only say that if 2, 3, 5 and 7 were all the primes then it is a prime and we have a contradiction. Proof by contradiction remains one of the most difficult things to grasp.

The primes do not stop at 7, but suppose they stop somewhere, at a number we shall call p. Then look at this number

$$2 \times 3 \times 5 \times 7 \times \ldots \times p + 1$$

where again we have all the primes multiplied together and we have then added 1. We follow the same argument. If the primes end at p, then we have a contradiction, for this number does not divide exactly by any prime and must therefore be prime; and it is certainly much bigger than p.

Close but brief reasoning of this sort is characteristic of the most able mathematical minds. We are enabled to see the whole of the reasoning in one go in our minds, not needing to follow pages of complicated symbols. Once we have 'seen' such a proof then it begins to appear obvious. That is after the event. Proofs of this sort open up short but entirely new channels in our thinking: we learn what a mind can do.

Despite the fact that there is no clear way of finding where the primes are, highly technical modern techniques allow us to estimate how many primes there are up to a certain level. The formula for the number of primes up to any number x is

$$\int_2^x \frac{dt}{\ln t}$$

If that does not convey much to you, it means that the estimate for the number of primes in the first million numbers is 41,606. We know, simply by finding them, that there are in fact 41,539. That is 67 out, a small fraction of one per cent.

There is something weird about an expression from the calculus being linked with primes, and there must be a deeply hidden reason for this. Perhaps as mystifying is the fact that it is only approximate, though always a good approximation.

There are people who hold that we should only study what is useful and 'relevant'. They also often hold that those matters seen as relevant will prove of interest to students of any age, and will 'motivate' (another OK word) them. This view is common among sociologists, particularly those who do not use experimental evidence.

The fact is that what interests our individual depends very much on the individual. It is what has connections for you. The question 'Do the primes go on for ever?' is one that intrigues many people, simply as a question. No one is going to need it for survival in life, but the opening up of methods of thought that are implied in this question could actually prove extremely 'useful' to a person in his or her development.

A living Chinese mathematician named Chen Jin-Run has spent most of his working life on Goldbach's conjecture, which we discuss elsewhere (p. 192). It says, or speculates, that every even number above four can be expressed as the sum of two primes. He has not solved it, but by now knows a great deal about prime numbers which most of us do not.

We might not expect politics to enter into prime numbers, but in certain countries anything is possible. While at one time he was highly esteemed for his efforts, under the changes wrought by the 'gang of four' it was felt that his work, lacking relevance, was not socially valuable. Before he was offered new work of more value, such as sweeping the streets, the political mood changed (to the discomfiture of Mao's wife and her colleagues) and he was rehabilitated.

It is not only in China that a mathematician, working happily in an ivory tower, can come unstuck. The United States National Security Agency has recently shown an interest in large prime numbers and wishes to classify part of the work, so that it may not be published.

This is a most extraordinary turn of events, yet the reason is simple. Prime numbers have come to be important in that most sensitive of areas, cryptography. As children we have all of us been fascinated by codes, and the passing of secret messages to one another. At the national level, the security agencies indulge their cloak and dagger fantasies in increasingly esoteric methods of encoding. Codes that regularly replace one letter by one particular other letter (or perhaps symbol, as in the Sherlock Holmes story 'The Dancing Men') are always easy to crack. During the last war, eminent mathematicians enabled us to read the most secret of German signals, and played a significant role in winning the war. Certain codes now depend on large prime numbers. Briefly the idea is this: if you take two very large prime numbers there is no problem in multiplying them together. Once you have the product, however, and give it to someone else, they may find it totally impossible to factorize it to get the original primes. This is not because we do not have a method. Eratosthenes knew the method. It is because sheer size eventually overwhelms even our computer. A large enough number (perhaps hundreds of digits) cannot be reduced to its factors by any means we know.

There are various methods of encoding, but consider just this. You have in your possession one very large prime. I have a variety of other large primes that I use as the basis for a coding procedure. I can take different primes at different times. If I transmit, in an open way, the product of your number and mine, you can determine mine, and others cannot, because they cannot crack the large number. That is not the whole story – there are endless sophistications – but the idea of a technique can be understood.

We have been trapped again. We claimed the mathematics we were doing was not 'relevant' and now find suddenly that it is a matter of national concern. Maybe the sociologists will let us study primes after all.

We seem to have wandered off from the basics into apparently airy issues, and then arrived back in wordly matters. We shall see this happening again in our chapter 'Mathematics in Action'.

Numeracy

This rather unhappy word was coined in the 1960s on the analogy of 'literacy'. As with many analogies it leads one astray. To define whether a person is literate (i.e. can read and write) is not totally straightforward, yet there is something reasonably tangible there. Mathematics does not have a central starting point like that, despite people's views on the basics. However, the word is there, and we must try to pin it down.

We have already revised our approach to the tables and can subsume them in the phrase 'a good acquaintance with the properties and relationships between the numbers up to 100'. We have not slung the tables out, we have rearranged and absorbed them. And our 'good acquaintance' is part of numeracy.

Next we have the 'four rules': addition, subtraction, multiplication and division, applied for the moment to whole numbers. Calculation is so large an issue that we must devote a chapter to it. For now let us accept that people need to know how to do these operations. No one will deny this; the issue is how far we push them, in particular now that we have pocket calculators.

From there the issue becomes less clear, but let us look at several broad areas. We take percentages, large numbers, estimation and interpreting tables of numbers, and look at them one by one.

The narrowest, in a sense, is percentages, yet it is a source of many misunderstandings among the public at large, and an area where there are those who wish to persuade us of things for their own commercial or political ends.

The reason for using percentages is sensible enough. Comparisons are always best made if you have a common baseline. In percentages we choose a baseline of 100. A simple problem illustrates this.

'One shopkeeper buys an article for £5 and sells it for £7. Another buys another type of article at £9 and sells it at £12.50. Who is making the bigger profit?'

The question is deliberately ambiguous. If each sold one of each, the second shopkeeper makes £3.50 profit and the first £2 and there is no doubt who makes more. Yet we have a basic feeling that this is not how we should measure it. It is normal to look at the proportional increase they make on what they themselves buy. If both invested the same amount of money in their goods and sold them all, we could then have a different and perhaps more sensible idea as to who was making most.

Our next step is to find a sum they might both spend. The smallest amount that allows both to buy an exact number of articles is £45.

Shopkeeper 1	9 articles at £5 each
	Profit $9 \times £2 = £18$
Shopkeeper 2	5 articles at £9 each
	Profit $5 \times £3.50 = £17.50$

and we come to the conclusion that the first shopkeeper has a slightly higher 'mark-up' than the second. A reasonable approach for comparing two shops, but it will become more and more troublesome as we introduce more shops. The standardization comes in with using £100 investment as the basis for everyone. For the two shops we have chosen £100 does not buy an exact number of articles, but no matter.

Shopkeeper 1	buys 100 articles for £500
	Profit £200
Shopkeeper 2	buys 100 articles for £900
	Profit £350

The first shopkeeper makes £200 profit on £500, which is £40 on every £100. The second shopkeeper makes £350 on £900, which is £350 divided by 9 or approximately £38.89 on every £100. The phrase 'on every' is the key. We are comparing everyone on what they make on £100. That is their percentage profit. Our first shop is making 40% profit and the second 38·89%.

There are traps even here. We have worked on the profit made on every £100 spent on stock. Our baseline is what he pays for it. Some shops suggest that if they buy at £100 and sell at £150 then their profit is not 50% but $33\frac{1}{3}$% (equivalent to the fraction $\frac{1}{3}$). They say (correctly) that only a third of what they sell it for is profit. It is simply a way of making the profit look less. Our starting point is when we buy and that should be our baseline. It is not only more honest but has the sensible result that an 80% profit yields twice as much money as a 40% profit. With sale price as basic, that would not be so.

The advent of VAT and an increasing tendency to negotiate wage rises on a percentage basis rather than as a straight rise has thrust percentages down our throats. With VAT we do not always need a clear understanding of percentage to sort it out. There are tables in which we can read off the amount of tax. None the less it is a relatively simple example to illustrate how we go about percentages.

If you have a bill, basic £30 but subject to tax, you may find some working like this:

£30
3.00
1.50
£34.50

This is because 10% is 1/10 and rather easy to find. Even if the amount were more awkward, like £32.40, we can still get 10% immediately as £3.24. What the bill then shows is another 5%, calculated as half the £3 and also added on. So we have added 10%, according to a simple rule, and then another 5%. All well and good if we have simple percentages.

There is a slightly more complicated way of looking at it, but one that is much more generally applicable, and easy since the advent of the pocket calculator. We know that 15% is 15/100 or 0·15 of the amount. We must add it to the amount and there are two ways of doing this. As before, take the £30 and add to it, £30 × 0·15 (which we do on a calculator if our mental arithmetic is not up to it). We get £4.50, and with some relief are back at the same answer, £34.50.

We added the money after doing the percentages. Why not add the percentage before doing the calculation? Instead of 0·15 we get 1·15. This simply means one and 15/100, or the whole amount with 15% added.

So: £30 × 1·15 = £34.50.

We have reached a simple and general rule. We could have learnt it at the beginning, but rules without reasons are very unsatisfying to the intelligent person. We can now mop up lots of examples with no bother.

Increase: £32.40 by 15% . . . 32·4 × 1·15 . . . £37.26
£83.72 by 12% . . . 83·7 × 1·12 . . . £93.77
£783.78 by 11% . . . 783.78 × 1·11 . . . £870

Life is better with a calculator.

Most prices seem to be increasing all the time, but we should look at discounts. Take a simple one to start with. 'If there is a discount of 20% on jeans at £15, what is the new price?

20% is 20/100 or one fifth. A fifth of £15 is £3 so the new price is £12. Simple enough with these numbers. This time jump straight to our percentages. 20% is 0·20. Decreasing by 20% means multiplying by 0·80 (or 1·00 – 0·20). We are in effect finding 80%.

Decrease: £32.76 by 20% . . . 32·76 × 0·80 . . . £26.21
£783.34 by 11% . . . 783·34 × 0·89 . . . £697.17

The basic idea is very easy, but in the past the difficulty of the calculation has obscured this.

A percentage increase in our pay has now become easy to calculate. We can readily find the difference in actual cash between a 4·3% rise and a 7·8% rise. Before the calculator, few wished to undertake the calculations. Of course, it may be that percentage rises are not the best bet for everyone. As a union leader once remarked of percentage rises, 'They give most to them that need it least, and least to them that need it most'. Quite apart from the neat balance of the sentence, and its seventeen successive monosyllables, the speaker was well up in numeracy.

A less numerate remark came from a spokesman of one of our most prestigious auctioneering companies, who commented when they increased their percentage 'take' that it was the first time they had done so despite years of inflation. If you do not 'get' it, look in the Appendix (p. 230).

The basic image for percentages, however difficult the numbers, is of movement up or down from a baseline, which we choose to call 100.

As children we find a fascination in thousands, millions, billions and so on. It is associated with our feelings about the numbers going on for ever. Yet it is difficult to get a 'feel' for a really large number. Starting small, therefore, it seems that the largest number most of us can grasp without actually counting is five. Beyond that, however quickly we do it, there is some mental arithmetic involved. We make use of our place value system, however, as a form of ladder to higher numbers. Start by asking what sort of collections normally come in numbers less than 10. Fingers on a hand, people in a car, rooms in a house . . . and so on. Next we look at those things that normally turn up as two digit numbers (that is between 10 and 100). The lines on a page, the children in a class, houses in a road, years in our life – again we may find examples outside the range, but it is a feel for size of number that we are after. We move on to three digit collections – the words on a page, children in a school, the minutes we sleep every day. Again they may spill over, but they are, as it were, a level up from the previous collections. We move into the thousands – perhaps grains in a packet of rice, words in a chapter (not too long a chapter), a smallish football crowd, and so on.

Each of these is an everyday example, familiar to most of us, and it is a good exercise to extend the range of these examples. It seems manageable into the thousands and rather more difficult beyond that. Think of our levels rather like those of a several storied house (at the moment four). Our big collections are on the top floor, called in England the third floor.

3rd floor	10^3+	grains of rice
2nd floor	10^2+	words on a page
1st floor	10^1+	years in our life
ground floor	10^0+	rooms in a house

We have started a notation for moving up. 10^1 simply means 10. The next floor is 10^2 – this means 10×10 or 100. 10^3 is 1000.

Pause a moment over the floor at ground level. We have put a zero there because it fitted the pattern and the way in which we use 0 for the ground floor in some lifts. We shall say no more than that 10^0+ is in fact the numbers 1–9; on that floor we were putting single digit numbers.

The next numbers we shall look at are in the range 10^4 upwards to 10^5. In other words they are in tens of thousands – big football crowds, words in a book, people in a smallish town, and so on. Once we get above that ceiling we are into hundreds of thousands and examples are a little more difficult to find. The salaries of 'top' people? A crowd at a really big pop festival – and what about the grains in a kilo of sugar?

If we edge over the million mark (10^6) we know that various national economic issues are expressed in millions (or many levels up from that) but to grasp even one million is asking quite a lot. There is one example that may help. Imagine the area covered by a tennis court and its surrounds, say at a professional indoor event. It is a size we are familiar with and can hold in our minds. Can we guess what would be a millionth of that area? Would it be too small to see, or larger than this page? When you have guessed look back to our hundred square. About half of this area (the numbers up to 50), blown up a million times, would give you a decent area to lay out a court with plenty of room to run round the edges. The reason is not too difficult. Measure the half table square and work out one thousand times each side. You will have the right sort of playing area. And on it you could put 1000 rows of 1000 half squares. A thousand times a thousand is a million.

Many of us think of the next level as a billion. In fact a billion is a number of levels up. In the same way that the Americans count the floors in buildings differently – counting the ground floor as one and not zero – so they put billion at a different level from us. They call one thousand million a billion, which is three levels up from a million, or 10^9. In Britain the billion is a million million; that is six floors up at 10^{12}. Negotiators of national debts, or the purchasers of certain death-dealing machines, need to know this.

We can see these numbers quoted in the media, but cannot hope to grasp them; we merely know it will cost us quite a lot each, even if there are fifty million of us to contribute.

It has become scientific practice to avoid writing long series of digits before the decimal point. They write numbers as if they were all 'ground floor' numbers between 1 and 10, and then indicate which floor we need to take the lift to. The sun is 93,000,000 miles away so we write it as $9{\cdot}3 \times 10^7$

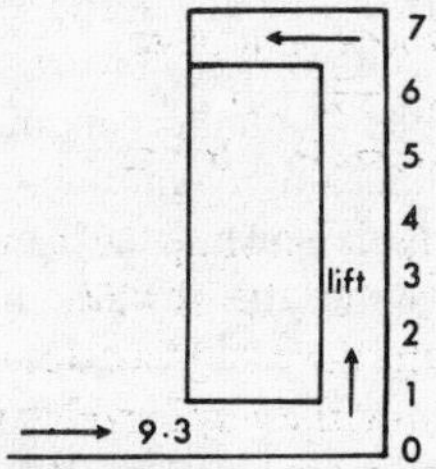

Orders of magnitude

Once we are into astronomy we really lose all sense of reality. With light travelling at about 670,000,000 mph and taking a great number of years to get almost anywhere outside the solar system, we do need rather large numbers. We have also gone a little beyond the realm of the morning paper, which mentions billions (American style) but calls it a day there. Nor should we expect numeracy to extend quite this far.

However, we might expect a degree of numeracy beyond that recently broadcast on a local station. It was said that a local organization in a single borough, wishing to help unemployment, was aiming to get sixty million pounds and had so far got seventeen thousand. In a single borough sixty million is in fact rather a lot to collect for a relatively small enterprise. Sixty thousand is more the mark. The confusion of thousands and millions is not encouraging among a reasonably intelligent group. They do differ by quite a lot.

At one time few of us really needed the big numbers but two recent developments mean that the ordinary citizen needs to come to terms with them. One is the size of numbers quoted in money terms in our papers; we have come gradually to adopt the American billion, 10^9, and need to know what it is. For a population of fifty million it is £20 a head. That a least gives us a feel for it.

The other reason we need it is because our pocket calculators slip into the scientific way of writing numbers when they get too big for the display screen. Most of them will take eight figure numbers (tens of millions) in their display, but go higher than that, and they will print something like 1·982437 9 and the 9 means we are above the ninth ceiling on the tenth floor.

Before we leave large numbers let us emphasize that the steps from floor to floor are not equal jumps up the number line, they are equal multiplications, and this makes an enormous difference. An increase of a floor multiplies by ten. It takes us, again in technical jargon, into the next 'order of magnitude'.

We cannot claim to be numerate without another skill that needs a 'feel for number'. This is the power to estimate rather than calculate, and in some ways this is the more important. In our last section on large numbers we have touched upon it. Placing a number in an order of magnitude is a rough process; the everyday examples of numbers of different sizes that we gave were all designed to give a rough idea of magnitude.

The ability to estimate is of regular value in our everyday life. When we go to do the week's shopping we need to know whether a £10 note will cover it, or whether we need a cheque, and if we take a cheque book whether the amount is going to exceed the £50 limit.

If we plan a day trip to the seaside, it matters to most of us how much it is going to cost – but we do not expect to reach an answer such as £21.43 before we start. We know roughly how far it is, roughly how many miles per gallon our car does, roughly how much lunch will cost us, and so on. We are likely to come to a conclusion such as 'a bit over twenty quid' and that is good enough. If our account at the end of the day is £21.43 we feel we made a good job of our estimate. If it is nearly thirty pounds we ask ourselves where we went wrong.

School books at one time were keen on problems about carpeting or papering a room. The detail on doors, windows, bays and so on was all very precise, and so had our answer to be. Papering a room is a practical activity but we do not set about it that way. Very limited experience of papering gives you a picture in your mind of the area covered by one roll. Looking round the room tells you that you need seven or eight rolls. Most shops are obliging enough to sell you eight and let you bring one back if you don't need it.

The calculator, as we often emphasize, is a very valuable tool. It can be relied upon to be accurate in what you ask it to do, but there is a danger in becoming too reliant. If you take an answer blindly without thought you can really be in trouble. Not because it has calculated wrongly but because you have not entered what you believed you had entered. Suppose you believe you have entered 353 divided by 24 and the answer comes out 2·208333. Some people are impressed by that string of decimals; it simply must be accurate! But if we are to travel 350 odd miles at 24 mpg we are not going to use just over two gallons.

We need an expectation of a particular answer before we start. This needs confidence with numbers. Be prepared to push them around a little. The aim is to do a calculation somewhere near where we want, something that we can do in our head. Dividing by 25 is not too bad; there are four every hundred. So in 350 we have three lots of four and a couple over. Fourteen. That is a reasonable amount of petrol and we expect the answer to be a bit more, because we do less than 25 mpg and need to go a little more than 350. Try the calculator again: 14·708333. Forget all the decimals. Reckon on 15 gallons.

How did we go wrong? If we play around on the calculator we find that we can get the 2·2 answer from 53 divided by 24. We did not notice we had not entered the first 3.

That reminds us . . . mental arithmetic. Not the timed test of school days, but a sensible level of working in your head. It is reasonable to hope that intelligent adults can work 350 divided by 25 in their heads. Probably it is near the upper limit of what it is worth bothering with. It goes beyond the table facts but does not involve us in becoming calculating prodigies. That is part of numeracy.

Much numerical information is presented in tabular form. One test for the numerate is simply whether such a table raises questions in your mind. The table of numbers we show is from some years ago, but the truth of the numbers is not the issue. Read them through and see what features occur to you. Certain numbers stand out, perhaps most obviously the zeros. So we ask ourselves why there should be zeros in those three positions . . . and why not in a fourth place? It is most productive if two or more people discuss the figures and argue about them.

Teacher ages

Age	Men		Women	
	Heads	*Others*	*Heads*	*Others*
20–24	0	90	0	234
25–34	2	250	3	580
35–44	13	143	12	197
45–54	12	94	19	144
55–64	19	83	13	66
65+	1	12	0	15
Totals	47	672	47	1236

Heads	94
Others	1908

The process of discussing numerical information seldom involves much calculation, but it does involve the feel for number that is an essential feature of numeracy. We might comment on and wonder why there are 'well over twice' as many women as men in the 25–34 age group. It does not forward the argument to know that it is 2·32 times as many.

The information presented here is neutral; it does not seek to persuade, although from the numbers may arise all sorts of far from neutral issues. The many persuaders in our society not only doctor the figures but supply their interpretation. The ability to interpret numbers presented like this is a skill needed by the ordinary citizen if he or she is to make informed choices. We discuss the table further in the Appendix (pp. 230–1).

In this chapter we have tried to reinterpret the basics and give a slightly wider meaning than is common. In fact, we could say that the whole of this book is what the responsible citizen should know about mathematics, but perhaps we should start with more limited aims.

2

Some practicalities

This chapter does not seek to say what mathematics is, but something of what it does. Most of the issues are without deep intellectual content; they are rather those things which make the wheels go round. They are also something of a random selection. Numerical matters (and most of these are to do with number) touch us in many areas of our life, and it would be a very extensive task to list them, let alone describe them.

Money matters

Buying and selling of all sorts has many traps, and detailed arithmetical knowledge is needed to be sure that what one is doing is right. The question is, to how much trouble are you prepared to go to get a good deal? This is a question that would be answered by each of us according to temperament; it could itself be costed – for our time is money. Some levels are not worth while, though who is to say that you should not engage in them? There are people so distrustful of machines that they will do the adding up of a shopping bill produced by the till at the supermarket. The bill needs checking, in that it is worth checking that each item is fed in once and once only, and that the correct price is entered. In doing this you are checking a person (and perhaps the store's policy) not the machine. It will add correctly what it has been told to add.

Odd-shaped boxes containing non-standard amounts of material, not always filling the box, at prices in far from round numbers, are designed to conceal the rate you are paying. Again, it is not worth a lot of effort; simply buy those things that are unit priced (i.e. the price per pound or kilo is quoted) and press for government action that all goods should be so priced.

As we move up the spending scale from the weekly shop it matters more. Buying a car is a reasonably sized transaction for most, and it bristles with sums. It is always easy in money matters to overemphasize one single issue, and for a car it is usually petrol consumption. With petrol as expensive as it is, it matters, but it is not as central an issue as some people believe. The motoring organizations can tell you some overall statistical facts about running a car and break down the costs of tax, insurance, garaging, depreciation, loss of capital investment, loss on hire purchase, repairs and so on – until we begin to wonder if petrol matters at all! It is not possible to show much detail here, but look at one or two salient facts of a single case.

Suppose you have £4000. Buy a car with it and it will probably drop to £3000 value in a year. Invest at 8% and you make £320 interest. Difference: £1320. If you travel that year 9000 miles at 30 mpg, that uses 300 gallons. At £1.70 per gallon that is £510.

The figures are rough and you can argue with them, but it makes the point that depreciation and loss on capital may matter a lot more than petrol – and we have not begun to estimate other costs. Some rough idea of what matters in overall costs, of how things break down, is important to everyone. The same is true of many other consumer durables – all the various mechanical servants that increasingly fill our dwellings. Try to break down their cost, using simple numbers, making rough estimates. In fact we move some distance away from exactness precisely in order to get a 'feel' for the numerical implications.

The next stage up is house buying, for most people the biggest personal financial deal they become involved with. This links in often with questions of one's income, life assurances against a mortgage and a range of matters that are high finances to most of us. There comes a stage when we mostly need professional advice, from our solicitor or bank manager; and it is worth paying for advice when the deal is so big a one in terms of personal finance. The interesting thing is that the numbers do not ultimately rule. Most financial operations can be run in different ways, offering options, sometimes costing different amounts but spread differently through one's life. We then choose what we want to do, not just on the basis of which is cheapest. Many people believe that a numerical understanding removes freedom of choice. This is nonsense. Numbers will tell you what things cost, not what you should do. 'Who shall be master?' as Humpty Dumpty remarked.

If you do have capital, and wish to invest, take advice – but not from a mathematician! In a sense money and trade calculations are a part of mathematics, but they are certainly not what mathematics is about, nor are mathematicians often either interested or competent. The Greeks made a distinction between arithmetica and logistica. The first was the study of number, in particular of the whole numbers and their relationships, which we shall look at in our next chapter. Logistica was the calculation used by tradesmen – and a trifle lower class.

Health

The gradual intrusion of mathematical techniques into various areas has been an important feature of this century. The relationship between the experience of a professional medical man and the widespread gathering of statistics is a fascinating one, particularly in diagnosis. An illness may have a variety of symptoms, and the doctor who has seen many patients will recognize the symptoms pointing towards a particular malady. In drawing on his experience, he is in effect compiling statistics, but on a much smaller scale than is now possible. He does much more, if he is sensitive, in assessing the weights to be attached to each symptom for a particular patient. There is nothing personal about statistics. Yet they have their role. If we know that appendicitis (notoriously difficult at times to diagnose) has, say, three or four main symptoms and responses to tests, and if we know that one symptom occurs in 80% of appendicitis cases, another in 40% of cases and so on, we can get a probability based on extensive information. This does not tell us for sure whether the patient has appendicitis or not, but it can guide us. Experiments in diagnosis have shown that doctors plus computer-based help do rather better than doctors.

A real triumph of statistics lay in the establishment of the link between smoking and lung cancer. The resistance to this work came from various sources. The tobacco manufacturers could hardly be expected to view the research with scientific objectivity, and so it proved. Addicted tobacco smokers showed an equal resistance to belief; others, whose reactions were simply to statistics, objected on those grounds. Statistics do not, of course, offer proof, and the acceptance of that fact by those who use them has been seized upon by those who do not want to believe in a particular possibility. While statisticians may say there is eight times the risk of lung cancer if you smoke rather than if you do not, the objectors say, 'Well, I know someone who smokes like a chimney and is as fit as a fiddle, and someone else who died of lung cancer who never smoked'. Certainly, it can be so. The statement is not incompatible with the statistics and certainly does not refute them. How do we persuade such people that their argument is not an argument?

As we said, there is no proof, there is probability. The real danger in statistical work is to assume that numerical relations necessarily imply cause and effect. It could be that some physical or personality defect makes some people more inclined to lung cancer. If then separately we find this type of person is also much more likely to smoke, then the statistics do not imply that smoking causes cancer. Such a situation is simply a speculation, however, and there is no evidence of such a single cause to both. It is common sense to believe that drawing smoke down into your lungs is bad for them.

It is interesting to speculate upon how far record-keeping and analysis of a numerical sort, computer based, will alter the way in which we operate in health matters in the future. The notion that numbers could be introduced into medicine would have surprised and pleased the Pythagoreans, even if this approach is decidedly less mystical than theirs.

Graphs and the media

Though we may think of teachers as those charged with 'explaining', increasingly it is the mass media who explain, and whose very existence depends on their power to do so. If a good proportion of the population does not comprehend, they will fail. Regrettably, good though their presentations may be, they are not always honest. Generally speaking TV, radio and the press have points of view to express and will tailor the information to support their views. An increasing amount of material seems numerical in form as we stagger from economic crisis to economic crisis – much of this information is presented in pictorial form, because the visual impact is for many people the most potent in forming their views.

Suppose we are concerned with a falling pound. Look at our two diagrams, and decide which represents the more serious situation. Used as we are to interpreting graphs, we see the first as a steeper fall to a lower level than the second. In fact, in the graphs as shown, no such conclusion can be reached, for the axes are not labelled. In the first we should ask if the fall is really steep and the level really low. How steep the line appears depends simply on the scale used on the horizontal axis.

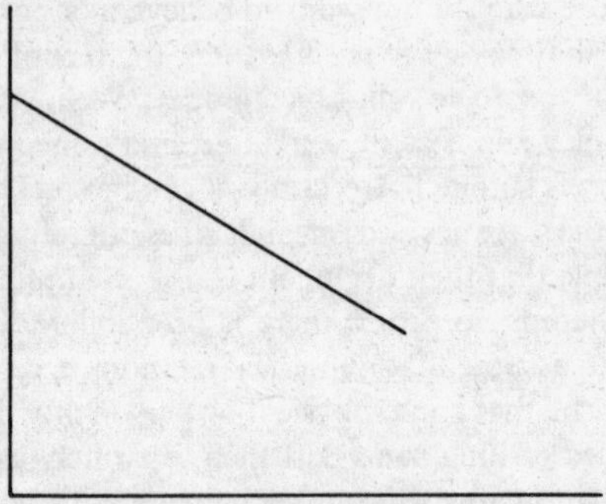

The falling pound (1)

If, in fact, the units in the first diagram are five times as large as those in the second, we would suddenly find that the second, when the scales are matched, was steeper than the first. Also the apparent depths to which it has fallen in the first depends on whether the horizontal axis is at zero level. If it is at a much higher level, whereas the second diagram does run at 'ground' level, then again we may find the second has fallen lower than the first. These two issues – what scales you use, and where you put the zeros – allow a great deal of misleading views to be conveyed. Look carefully for the numbers on the axes, and use your common sense.

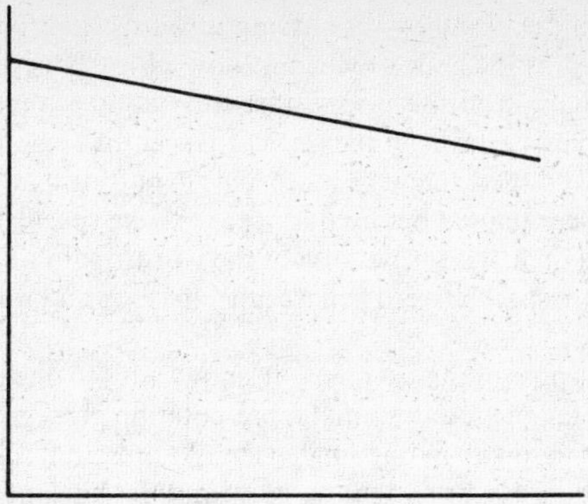

The falling pound (2)

On the one hand we have seen more and more explanation carried by graphs and pictorial representation rather than prose. On the other we have seen a most interesting penetration of the language by images drawn from graphs. Let us look at one or two examples.

> 'After a steady period of about a year, house prices have again begun to rise'.

This is a perfectly clear language statement, easily understood by most people. Let us do the reverse process, and instead of trying to make some language statements from a graph, turn this statement back into a graph. Our diagram shows the completed picture, but there are stages in arriving at it.

We start by drawing a flat line 'about a year' long. Then we show a rising line. We have made is straight, and that could certainly be argued about.

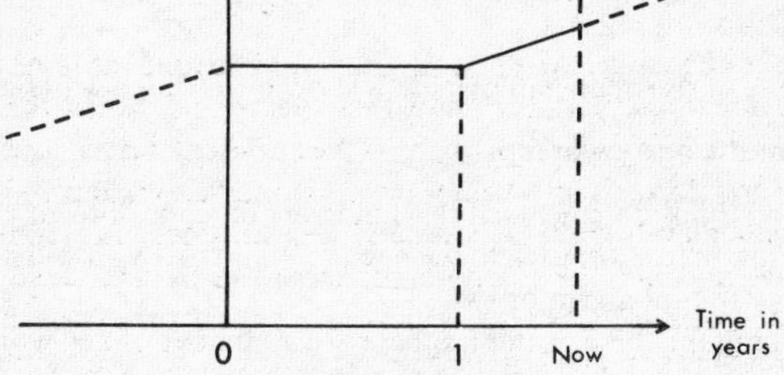

House prices

Perhaps we should have started it curving upwards slightly. Then, implicit, though not stated, is the issue of when is 'now'. Clearly beyond the year flat period, for we have seen the prices start to rise. So we mark 'now' in an indefinite position. We have put the vertical axis at the beginning of that year; that is arbitrary and there might be more logic in putting it at 'now', even if we do not know where that is. Before our year we have speculated (dotted line) that it was rising then. This is because of the word 'again'. Beyond 'now' the dotted line is more speculative. Then we label the axes and we have a neat and tidy graph.

As to the purpose of this exercise – it simply adds to one's understanding. The marriage of language and spatial representation gives greater insight and depth of understanding than one single view. At present it is believed that the two hemispheres of our brain tend to control different functions. The left is normally dominant and controls language, numerical work and most analytic thinking; the right gives us our spatial perceptions and may be a larger factor in creativity and problem-solving. The constant use, then, of spatial interpretations can be of value in problems less mundane than that we have taken as an example. There can in an individual be a marked imbalance in these two complementary abilities.

It would be wrong to harp too much on the ways that we may be deceived, for much information is offered as objectively as the presenter can manage, yet this use of language in place of graphs can certainly be misused. Look at these three political promises:

> 'We will cut prices at a stroke.'
> 'We will cut price rises at a stroke.'
> 'We will cut the rate of price rises at a stroke.'

Which way does your vote go? Can you draw the pictures?

Suppose you are concerned about presenting your party's performance on unemployment. Regrettably you find that you cannot actually claim that unemployment is falling, for it demonstrably is not. So we next ask if perhaps it is not rising as fast as it was. (Is there a picture in your mind?) Regrettably again, we find it is not so. We make a last and final effort and announce proudly:

> 'The rate of rise of unemployment is showing a tendency to slacken.'

Cheers from the party workers.

Lest you think statements of this kind are not made, look at your 'heavy' newspaper tomorrow, or listen to the news on radio or TV. It will not take you long any day to find a gem or two.

In looking for these examples there are a number of key words, such as 'rise', 'fall', 'steady', 'increase', 'rate' and so on. It is an amusing exercise to compile a list of them and then sort through a single day's newspapers for sentences involving them. Your attempts to express them as we did the house prices may or may not be successful, and that may be a lack of skill on your part – but is more likely to mean that the sentence does not in fact have a meaning. It is a sobering exercise.

Sport

Once number has entered any human activity it seems to have an almost infinite capacity for growth. Games are based on numbers, and we shall see something more of them in our chapter on probability. Mostly the issue starts simply, but the sport statistician is often a monomaniac who can always think of new things to calcuate.

Games generally involve scoring more points than your opponent, though some, like golf and certain athletic events, end up with the lower score winning. Mostly we start with whole numbers, and even if games like tennis or bridge use, for no apparent purpose, numbers that could easily be divided by five or ten, we normally start with counting numbers. However, the desire of some people to use the four rules soon overwhelms us.

The performance of individual batsmen, be it at cricket or baseball, are expressed in 'averages' calculated to at least two decimal places. Clearly they have some meaning. The great Don Bradman had in his test career an average of 99·96; not encouraging to bowlers. Some other well-known cricketers were held in high esteem for an average of half that. Were they only half as good? Presumably, if averages are to mean anything. Naturally people then begin to argue about the conditions in which they were made and how Bradman compared with others on a 'sticky' wicket.

To argument on sport there is no end. Before we leave cricket, how about this from a statistician?

'75 is the lowest score that Cowdrey has never made'

Just one of these nuggets that enhance the sport.

Football league championships, too were sometimes decided by the second decimal place – one of the few times the ordinary person could be induced to pay any attention to decimals. If the points (at that time two for a win and one for a draw) were equal for two teams, their 'goal average' was calculated. You divided the goals they scored by the goals scored against them; the higher the result, the better for the team. This has now been replaced by 'goal difference'. At least we now have no trouble with the 'damned dots' and we can stick with the counting numbers.

Arguments arise, both in sport and elsewhere, about the 'fairness' of different methods, as if there were an external means of judging. Is the 'difference' method fairer than the 'average'? There is no answer. There is a corresponding situation in wage claims: is it fairer to have flat rate rises, or proportional rises, expressed in percentages? Our union leader of the last chapter had no doubt about the inequity of percentage rises.

Some of the numbers that arise in sports lead us into probabilities, a topic we shall return to later. What is the advantage that a home team has, irrespective of sport? Clearly we can simply take the proportion of home to away wins over a season and we have a measure. It is possible with a great deal of work to get quite good forecasting of results using such measures, only in a more sophisticated form. It is even possible to have steady wins on football pools over a longish while – but the big ones come in the weeks with bizarre results that you will not forecast.

Number has also penetrated several sports in establishing 'rankings' of players. This is an interesting exercise, and is particularly prevalent in person against person type games. Both in tennis and in chess, players at many levels can be given a grading. In tennis, computer rankings mainly apply to people on the international circuit, constantly playing one another. In chess the Elo rating (named after a professor) attaches a ranking not only to grand masters, but to those at many lower levels, so that a fairly ordinary player can get a grading on the same scale as the world champion, even if he does get a much lower number.

Trigonometry: how steep is a hill?

It has to be a nice even hill, climbing steadily like this

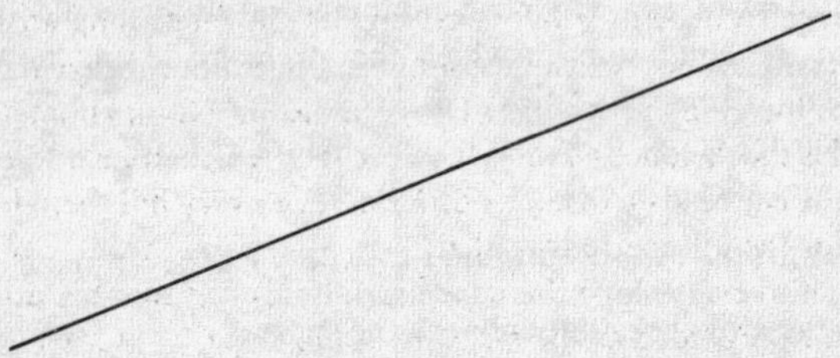

and we imagine that we are provided with a map with contour lines on it. There are many ways we can describe the slope. We could, for instance, say how far we had risen when we had driven, or walked, one kilometre:

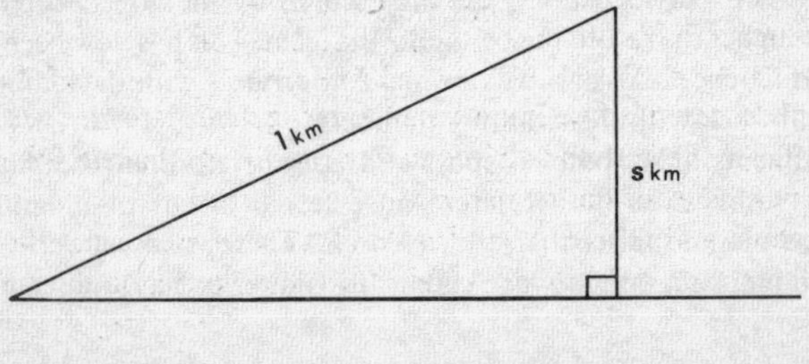

We've called it **s** km – it might be any value. For a pretty steep hill it might be, say, 0·5 km. This number 0·5 is *one* measure of how steep the hill is. Obviously, when there is no hill and we are travelling on the level, **s** is zero. For a vertical hill, a cliff-face, **s** will be 1. Any number between 0 and 1 gives a definite, unique hill.

Once we attach a number to a notion, such as steepness, we have a *measure*. But we could have decided to use another measure. It would be equally natural to find how far we had travelled on the map.

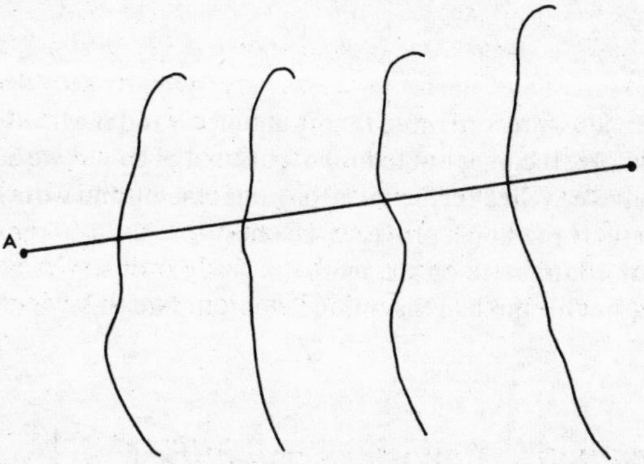

The road travels from A to B through rising contour lines. We can find the direct distance from A to B by measuring on the map using an appropriate scale. But it will not be the distance shown on our car's mileage recorded. The car has been travelling up a slope. The map distance for a kilometre travelled on the road is the distance **c** km. AB on the map might appear to be, say, 0.86 km, while the car reads a distance of 1 km. For every hill there is a different value of c for the kilometre travelled.

Again, if to every slope there is one and only one number attached, that number can be regarded as the measure of the slope, or at least *a* measure of the slope. So we may write the same slope as

	s	**c**
slope	0·5	0·86

Either will do, provided we know which is being used – whether we are being told how far we have risen in a kilometre drive or whether we are being told how far we have gone on the map.

We could go on inventing all sorts of ways of measuring the slope. It might, for instance, be sensible to make the measurement on the map simple, and not that on the road. When we are planning from the map we might decide to measure how far we have risen for 1 kilometre on the map like this:

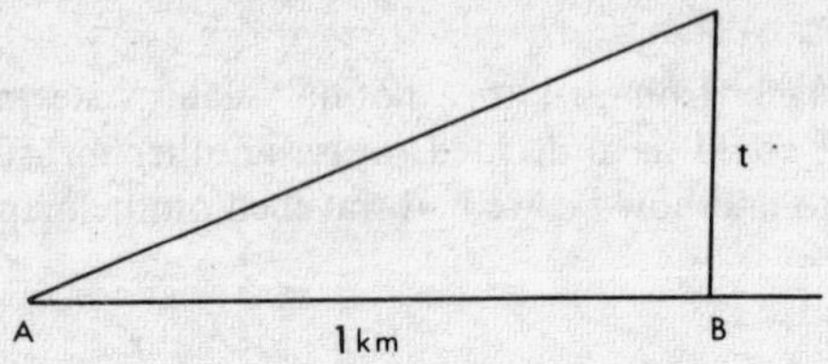

Our car will obviously record more than 1 kilometre and it will also have risen further than before. It comes out to be 0·57, instead of the 0·5 we had for **s**. But although we have travelled further, it is the same slope – and we now could say that 0·57 represents the slope, provided we know that the 0·57 represented the height risen for 1 kilometre on the map.

We could put all the lengths on one diagram like this:

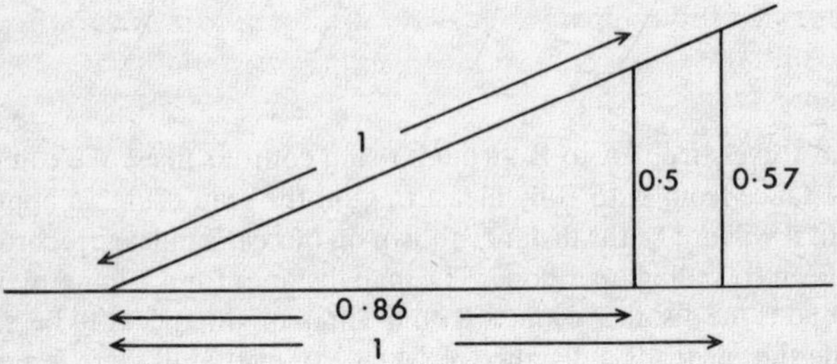

The general point about measures is that any number that attaches uniquely to one position of the thing measured is satisfactory as a measure. All three of the numbers 0·5, 0·86 and 0·57 are measures of the slope of that particular hill. Any will do, provided that when comparing slopes you use the same measure for each.

The particular slope we have been dealing with was in fact a 30° slope, which would be pretty steep for a car, and a not inconsiderable one for walking. What of the three separate numbers we have associated with it? They are our old friends (or possibly enemies) sine, cosine and tangent. Our now obsolete log tables, or their replacement the pocket caclulator, give us the values for any angle we like.

These three numbers allow us to do a wide range of calculations. Let us look at one or two. In the first diagram we are measuring a church tower. We have stood off 100 yds from the base, and using a clinometer find that the top of the tower is at an angle of 23°. Look at our map analogy. If we measure our distance on the map, not the hill, we are using tangent. Look up, or press the key for, tan 23°. It is 0·4245, though we may not want all those figures. So one on the map takes us 0·4245 up. Hence 100 yds on the map takes us 42·45 yds up. That is its height. Since we usually measure in feet, we multiply by 3 and get 127·35 ft, which we round off to 127 ft, since there is no guarantee that we measure the angles as accurately as the tangent.

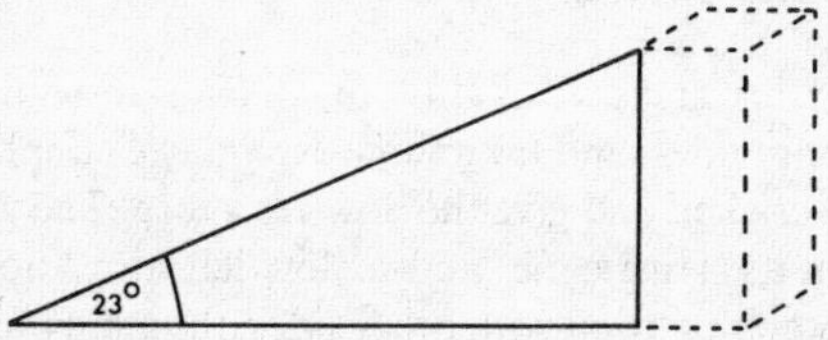

The church tower

Our second problem involves two men carrying a 20 ft ladder round a right angle turn in a corridor 7 ft wide. Can they get round the corner? Our diagram shows the ladder in the most critical position. We have 10 ft of ladder slanting across each part of the corridor – is the 7 ft width enough or not? This time our starting point is the 10 ft slope. Our base is what we are interested in and our analogy tells us it is the cosine. Now cosine 45° is 0·7071 and since we had not 1 but 10 along the slope, we need a base of 7·071 ft. It is just too small – but perhaps by bending or tilting the ladder we may manage.

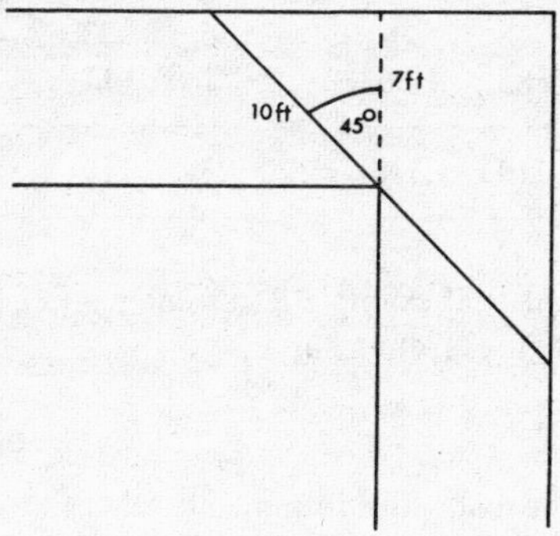

The ladder problem

Trigonometry has wide uses in surveying and such practical matters. The trigonometric ratios later turn out to have a deeper significance within mathematics itself, but for now we are concerned with practicalities. There is no mystery about trigonometry, it is simply about ratios in triangles. It only appears difficult if we worry about the complications of intricate diagrams with many lines. The underlying idea, as so often in mathematics, is accessible to all.

Traps for the unwary

Though there are other areas we might look at, let us instead end this chapter with some cautionary tales, and take some everyday examples where we may be misled. Back to cricket, and give the bowlers a turn. The village team has a fast bowler known as 'Hurricane' and a slow left arm known as 'Twister'. Though they enjoy competition with other sides, the main point of playing for each of them is to outdo the other. For the first four matches, Hurricane claims a better average every match than Twister, who is none too pleased and points to the fact that he is at least taking more wickets.

Hurricane is needled by this, and in the fifth and last match ruins the opposition (and Twister, his main opponent) with a devasting bowling spell. He even has the satisfaction of having yet again a better bowling average for the match. 'That'll show him!' Yet when we add up at the end of the season, who is on top? That is right – Twister. There is some justice after all.

Hurricane			**Twister**		
wkts	runs	aver.	wkts	runs	aver.
2	10	5	5	30	6
1	3	3	4	15	3·75
3	13	4·33	8	40	5
2	8	4	5	23	4·6
8	85	10·63	1	11	11
16	119	7·44	23	119	5·17

For our next statistical example let us enter into extramarital sex. We quote the following figures from a survey:

UK 1969	Men	Women
People (%) who've made love to someone other than their husband/wife since marriage	7	3
People who haven't	62	82

Presumably there were the usual percentages of 'don't knows'. (For our next question, 'What proportion are lying?') Let us look at the way statistics of this sort are arrived at. Imagine a country with some rather odd laws on marriage. Marriage is a limited term contract, entered into by every one at age 20 and ending for everyone at 40. Laws on fidelity are very strict. For the first ten years of marriage infidelity is punishable by a slow and very unpleasant death. Moreover, surveillance is absolute and inescapable. So no one errs. At the age of 30 adulthood is reached and adultery not merely encouraged but enforced. In consequence, when we make our survey we find the table for that country is as follows:

	Men	Women
People (%) who've made love to someone other than their husband/wife since marriage	50	50
People who haven't	50	50

You will observe that at least in this respect it is a non-sexist society. Yet the figures as presented tell us little about their actual marital or extramarital behaviour. Perhaps the same is true for the figures for the UK.

From time to time shady characters start 'pyramid' swindles of one sort or another. There are many variants, but a basic pattern is something like the following. Someone starts and gets two friends to recruit two more people each, who then recruit two more and so on. This establishes a pattern like this:

A	1
B C	2
D E F G	4
H I J K L M N O	8
	16
	32

and so on. The inducement to establish this pyramid is that, though on entering you have to send £1 to the person five levels above you (that is A, if you are at the level with 32 in them), then after A has been paid, B and C share the amount sent by the 64 line; you eventually reach the top and get £32 for your £1 investment. The only people likely to make a profit are the first 5 rows, who sent no one anything. The danger in joining is that we never recruit five levels down from where we entered. if we did, it would be a very large population. Like most schemes, it benefits those who devised it.

Finally, beware of the newspaper you read. One of our well known newspapers published the following scare headline: '50% of our pupils leave school without a single O level.' Disastrous. What a condemnation of our educational system. Now let us look at it objectively. The examinations at O level are designed to meet the needs of 20–25% of our population. Large enough numbers enter, particularly in the major subjects, to ensure fairly stable pass levels. It is *arranged* that only 20–25% of the ability range should pass. In consequence, it would be foolish even to enter someone who was halfway down the range. Even allowing for different performances in different subjects by particular pupils it is totally unremarkable that 50% of our pupils leave school with no O levels. Is it imagined that if we all worked hard enough we would all get some O levels? Do not believe it; the hardfaced examiners would shift the standard upwards.

Mathematics has its severely practical aspects and copes with many issues important in the day-to-day running of affairs. It is many-faceted, and to see it simply in terms of those places where we encounter it would be a mistake. Let us look further into it.

3

The story of calculation

Rather like 'The Story of Steam' it begins to look as if we can tell the whole story, from beginning to end. We may not be sure when it started, but we may be at the end just about now.

Pre counting

The development of number sense in a preschool child mirrors its development over millennia of prehistory in the development of the human race. It is this condensing of race experience over enormous stretches of time to a year or two of each new child's life that epitomizes our ability to learn, and is the root of our dominance over other species.

As far as we can tell, many species have some inbuilt notion of number. It is said that if one egg from three is removed from a bird's nest, the mother exhibits great anxiety; if there are four, or perhaps five, originally, the loss of one is not noticed. How much of this is specific research and how much folklore is difficult to say, yet consider our own innate perceptions. It seems (and we can try this for ourselves) that we can 'see' how many objects there are up to about five. Beyond that, however swift the process, if we see seven, we mentally split them, probably three and four, and do an addition sum. The process of 'seeing' without any calculations seems to stop at five.

If we attempt an experiment corresponding to the one with the birds, we find that we do have some grasp beyond five. Scatter some sweets on the table, turn away, and ask someone either to remove one or not remove one. Then test whether you know. It is not easy to conduct this experiment, since we are always tempted to count, but a brief glance by someone determined simply to take in the whole picture can work. Our ability to perceive number in this way may extend as high as ten, but if there were, say, thirty scattered sweets, it seems unlikely that anyone could decide whether one had been removed or not.

The next stage beyond just knowing how many there are is still pre-counting. In mathematics we refer to establishing a 'one-to-one correpondence'. We shall meet this notion again, and it is an important one. Once we had a number of possessions and began to organize ourselves into tribes, we had a need to keep track of things. The chief wished to know if everyone was present, the shepherd whether all his sheep had returned. It is quite difficult for us to know who is missing even in a small group. If a football team meets on a street corner, think of the problem of knowing who is missing without counting and listing; it does, of course, depend on whether the person missing is a powerful personality or not. Counting tells us if anyone is missing; who is missing is less easy. Yet we can find if someone is missing without counting. If we keep a number of pebbles, matching 'one-to-one' with people, by pairing off we do not know how many people there are, but we do know if they are all there.

If our sweets are chocolates in a large box we immediately know if one is missing. The hollow into which it goes is empty. We still do not know how many there are. The method is very effective. The early shepherd simply moved a pebble from one pile to another as each sheep entered the pen, and knew at the end if any were missing. Infant children play games, laying table for instance, where this prerequisite for counting is established.

Another one-to-one correspondence is the notches on a tally stick (from the French *tailler*, to cut), widely used in the Middle Ages to record number. If two merchants wanted a permanent record of a transaction, the stick was split lengthwise through the notches, each merchant retaining half of a matching pair. All in all, quite a remarkable number of numerical transactions can be managed without counting, let alone calculation.

Counting

Although we all learn to count very early on, the process is quite intricate. We need to learn the number names in our particular language. This gets mixed in with all the extensive language acquisition of the preschool child. He or she learns 'One, two, three . . .' but commonly gets the words, without always getting the order correct. With constant parental correction the child learns the number names in the right order up to, say, twenty; the fond parents sometimes then believe that their offspring can count, but this is not so. There are some stages yet to go.

We again use one-to-one correspondence, and we attach a number name to each object. Watch a young child trying to count a pile of books, say. The process starts well, and the first few books are counted, but the child's finger then begins to slip over the books, and the number names are said without being matched. The establishment of a firm matching takes some time in the child's development. The final step is an odd one, when we look at it carefully. The last number name we reach has then to be attached to the whole collection. This can be very muddling. If to the bottom book of the pile we say 'thirteen' we then have to shift to the idea that thirteen characterizes not that bottom book but the whole pile. This is not a trivial distinction; although as adults we are all familiar with the process and may not think much about it, it is significant, both in the basics of mathematics, and in the child's psychology.

Numerals

Spoken language precedes by far the written words in our history. In the child this procedure is also true. When he or she arrives at school an early task is to establish connections between spoken word and real objects, and then to link with the written word. A similar process is necessary in number. Number itself is abstract, and we shall shortly consider it at length. The spoken words for number, the number names, are learnt to facilitate counting. The written symbols for number, the numerals, form the third element as in language. A numeral then is 7 or 9; it may be written differently in some parts of the world. It is not a number any more than 'cat' is a cat. The distinction between the thing itself and the way in which we write it is an important one, yet few would draw this distinction between number and numeral.

Evidently drawing goes back beyond the written word. The wonderful artists of the Lascaux caves could not have written, yet it must have been from drawing that some symbols for words derived. The earliest symbols for numbers were simple marks or notches. They could have arisen very early in our history; perhaps in a sense they are the first form of writing, as opposed to drawing. Since the numbers are essentially abstract, so too will be the symbols representing them.

The Chinese, the oldest of our civilizations, started with tally marks, and used brushes and black paint. Their first five numbers were

1 11 111 1111 11111

but the difficulty of continuing with this (effectively one-to-one correspondence) led to the next numbers being written

$\overline{1}$ $\overline{11}$ $\overline{111}$ $\overline{1111}$

though later the arrangements were modified, and we have symbols such as ŧ for 7 and ∓ for 1000.

The Babylonians established the most effective of the ancient systems of numeration. Traces of their method of counting in sixties remain in our measurement of time, with sixty seconds in a minute and sixty minutes in an hour, as well as in our decision to divide a complete turn into 360 degrees.

Their *cuneiform* or wedge-shaped symbols were made with a special tool pressed into clay.

▼ was the symbol for one
▼▼ was two

and this pattern of repeating the sign, common to many early systems, continued up to nine, with symbols written in rows of three. They then turned the sign on its side

◀ represented the number ten

Had they pursued their earlier notation, it would have been possible to go up to ninty-nine. In fact, they wrote the tens in a string but stopped with five of them. This, together with the units, took them to fifty-nine. It was here that what we now think of as base sixty entered. They decided to turn the wedge again, back to the first position, and yet it now was meant to be sixty. In this they closely approached our present-day place value system. In this system

▼◀◀▼▼▼ meant $60+(2\times 10)+3$ or 83

This left a problem with numbers that had an exact number of sixties with no spare terms.

▼▼▼▼

This array might mean 4 or 63 or perhaps 122. It is similar to our problem if we had 8372 and were told there was a decimal point somewhere, but not where. They cleared the matter up in a very similar way, writing

▼⋟▼▼▼ for 63

with the sloping stabs separating off the sixties.

The Egyptians had a system more directly ten-based, arranging the units as did the Babylonians in rows of three, but using single vertical marks. For them ∩ was 10 and C was 100. As with other systems, the symbols were repeated with the assumption they were to be added.

Oddly, the Greeks, to whom we owe so much in our mathematical thought, used a far inferior system of numeration. They used their alphabet, supplemented with three extra symbols, to represent the numbers 1–9, 10–90 (in tens) and 100–900 (in hundreds). They could represent larger numbers by a system of slashes: ͵α meant 1000, not 1, and so on. Recording the numbers was simple enough but calculation was very difficult indeed.

The Romans made some improvements, and because of their widespread conquests their numerals, like their language, persisted for a long time. Yet they were not adequate for detailed computation either. They remain familiar to us and are still occasionally used, for instance when we wish to use different numerals for different subsections in a detailed report. It was basically a tens system, though new symbols were introduced alternately in multiples of five and of two, like this:

I	represents	1
V	,,	5
X	,,	10
L	,,	50
C	,,	100

and so on. The tens system was effectively broken down into fives and twos.

A comparison of these methods illustrates the features that may be present in a numeration system. They all use the repeated symbol to mean that they are to be added. That is a natural enough thing to do, but we shall see how helpful its abandonment was. Except for the Babylonian and Greek, new symbols had constantly to be introduced, even if less and less frequently as we went up the numbers. The Greeks used an inconvenient system of dashes to avoid using more symbols but the Babylonians had an idea not used elsewhere. They used the same symbol but made it mean something different according to where it was. This was the really significant difference in the Babylonian system, and one which is of great importance in our own system.

The Hindu-Arabic place value system

The efficiency of the way we write numbers is not apparent until we see it in the context of these number systems. It was developed by the Hindus and spread by the Arabs, who were great traders, and who added the sign for zero, a most important step.

It is worth analysing the characteristics of the way we write our numbers. For each of the first nine numbers we have a separate and different symbol. This is like the Greek system, but unlike those with repeating symbols, such as the Babylonian Egyptian or Roman. We then, effectively, make a bundle of ten and call it, not by a new name or symbol, but by the symbol for one. We teach this to young children, who collect ten sticks and make a single bundle. Not until we have collected ten bundles, and again made one large bundle, are we really moving into a properly based system. Marking a hundred as 1 and then a thousand as 1 means we go on for ever with the limited range of symbols we started with. This marks out our system from the others. The final move is what constitutes place value. To have a constant base (say 10) and even to have a limited set of symbols does not constitute place value. For that we imagine a separate grid into which we slot the symbols; our diagram indicates a track with spaces labelled, rather like the HTU we use in primary school. If we drop the 98 straight down it means ninety-eight thousand. Put the 8 up against the decimal point and it means 98. Move it along two places and we have ·98, just below one.

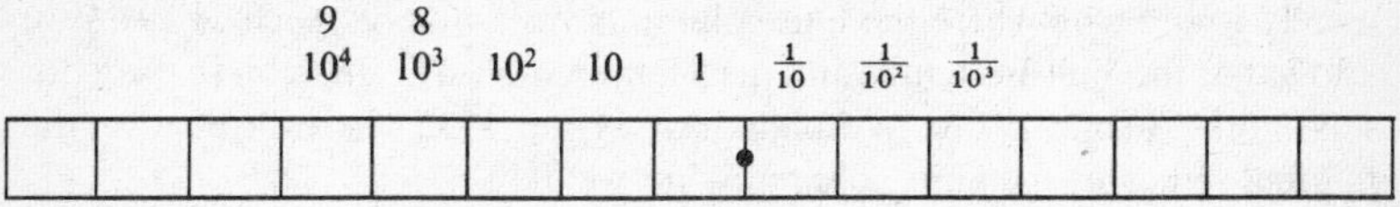

But back to where we had just reached ten. Using 1 to mean ten involves putting it in a certain place in our grid – but how do we show clearly which space? The use of zero as a place-holder was the important move. So we write, as we know, 10. The place-holder can, of course, come in the middle of a number: 503 simply means there is nothing in the tens column.

Numeration and calculation

The way we write a number is a human decision. We have seen how different the decisions were in various civilizations. The use of 10 as a base is quite arbitrary, and presumably arises through our having ten fingers. Any other base would allow us to write numbers, and ten may not actually be the most convenient. The symbols are as we choose; it is good luck that we have established a wider acceptance of one lot of symbols in number than is true in language. Number itself is not affected. Seven is always prime, no matter how we choose to write it.

Should we therefore attach much importance to how we write our numbers, and does it affect the world at all? It is interesting to speculate where our society would stand now without the introduction of our system. To ask what may seem to be a rather odd question, would we have motor-cars?

Technological advances in Greek and Roman civilizations were seriously hampered by an inability to calculate easily. Try to multiply MMCCXLII by MCXXV if you are in any doubt. Our use of the same symbol in different places means that we have only to know relatively few number facts, as we showed in discussing the tables. If we know $7 \times 3 = 21$ we are enabled to use it, wherever the 7 and the 3 are positioned in the grid.

The sort of calculation that Kepler did in the seventeenth century to determine the movements of the planets would not have been feasible in some systems of numeration. The developing economic nature of all human organizations has necessitated rapid calculation and an easily examined system of recording transactions; the tally stick no longer works.

The Romans were considerable engineers, but anything remotely resembling our present-day precision engineering demands measurement and calculation of a very precise nature. Another extension of the place value system solved this problem for us. Once the trick of the digits in adjacent columns being ten times each other (or looking at it the other way, tenths) is grasped, it is natural to go below the unit columns, labelling tenths, hundredths, thousandths . . . as small as we like. We introduce a decimal point merely to indicate where the unit column is. In a number such as 19·35 we know that the 3 means tenths and the 5 hundredths. The decision to use a dot was another arbitrary one. We might have written $1\bar{9}35$ where the bar over the number indicates the units columns – but we didn't.

It is odd that with place value well established for counting, we yet managed to produce a system for numbers below one which was not place value – the fractions. Certain fractions must arise from social situations where sharing is involved, and we cannot think of doing without, say, halves and thirds, but their further development, though having some mathematical interest, has been something of a disaster both for practical matters and for the education of children, who have not always enjoyed the 'hard fractions' offered them.

As in place value, a decision to write symbols in a certain way has led to considerable consequences in fractions, not all desirable. We have a mismatch in the convention used above and below one. It was a step forward in dealing with more precise measurement to use fractions, but not as simple a step as the more natural notation of decimals. With these we can be as accurate as we wish, use a consistent notation, and have the same process of computation above and below one. If we have, say, $1437 \times 0{\cdot}532$, the 7 and the 2 multiply together to give 14, though we have to know where to place it. The final breakthrough in this work achieved by the pocket calculator removes the need for huge swathes of wasted learning of computational techniques.

We are nearing an understanding of the question of whether we could build a car without place value. At least the issue has meaning.

Our world would look markedly different were it not for place value. An extraordinary situation, that a decision to record in a certain manner should have such consequences; there is no surprise that new discoveries and advances in technology should effect drastic change, yet there is something mystifying in changes brought about through a symbol system.

The abacus

This calculating device has the longest history of any. It dates back to the Babylonians and is still in use today, with little variation in form. Our diagram shows a simple abacus. The beads on the wire above the line represent five

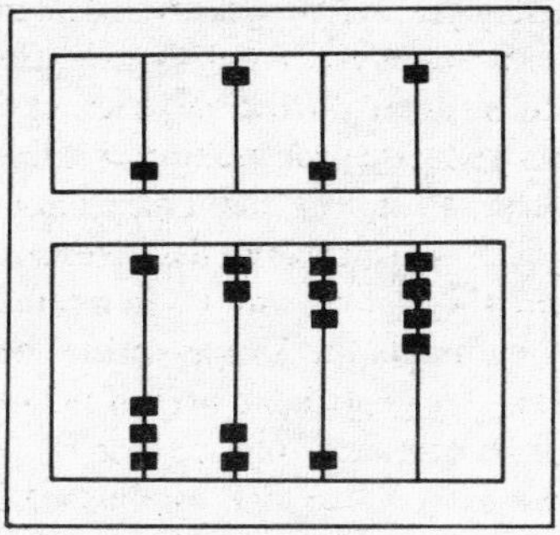

The abacus

units, those below represent units, giving a total of nine in every full column. The system was place value, and our diagram shows the number 6284. It is those beads which have been moved in to the central bar which record the number. In adding we use the 'spill over' method where once a column becomes full we empty it and add one to the next column, exactly as we 'carry' in pencil and paper work.

Practised users have developed skills, and by combining mental arithmetic and the use of the beads they have been able until recent times to work quicker even than some machines.

Pencil and paper algorithms

For most of us this is what computation (or perhaps mathematics) was seen to be about. We steadily made our way through the 'four rules' with the numbers getting worse and worse. If we mastered long division of decimals we might even penetrate the mysteries of the 'square root by the long method'. Pages of neat calculations, hopefully all ticked, encouraged us to believe that we were advancing in mathematics.

The development of printing was an enormous step forward in our history, as much for mathematics as for language, and permitted semi-permanent records of transactions in trade. The first printed arithmetic was published in Treviso in Italy in 1478. In England Robert Recorde (*c.* 1510–1558) was particularly influential, and his book *The Ground of Artes* was published around 1540 and set a style for generations of books on arithmetic.

Methods gradually became refined until reasonably standard approaches became stabilized. To learn a method by rote is of little value; only when it is given a full justification, and the reasons for the steps established, can anyone benefit. Slightly different routines have attracted strong adherents. For instance, look at these:

$$\begin{array}{r} 236 \\ 83 \\ \hline 708 \\ 18880 \\ \hline 19588 \\ \hline \end{array} \qquad \begin{array}{r} 236 \\ 83 \\ \hline 18880 \\ 708 \\ \hline 19588 \\ \hline \end{array}$$

In one we decided to multiply by the 3 first, in the other by the 8. It simply does not matter; what matters is the understanding of place value that assures us that the 8 is really 80. Now examine a long division:

```
      89
73 ) 6497
     584
     ---
      657
      657
      ===
```

If you remember it as a series of routines, say '7's into 64 might go 9, try 9, then if not right try 8' . . . you are seeking to become a machine.

Consider what the problem really is. We are asked to take 73 away from 6497 as often as possible. This could be done once at a time, keeping a tally, but it would be a lengthy process. So we take it away in large chunks. The 8 that we arrive at is in fact 80 – and it is possible to remove 80 lots of 73 but not 90. It is worth re-examining our knowledge of any of these processes to see if we can make sense of them.

Admirable though our place value system is, and important though it has been in our technological advance, in some ways it tends to obscure understanding. The very routine of saying, for instance, $7 \times 8 = 56$ in the middle of a computation can hide the size of the numbers, which may in fact be $70 \times 800 = 56000$. The use of the same symbol, that great step forward in notation, also conceals the fact that the symbol is not the same number.

In arguing that now we waste a great deal of time in fruitless learning of these skills, we do not want to deny their importance at one time. To be a 'scholar' at one time meant to be able to read, write and figure; those with these skills justly enjoyed high status. The modern world simply demands different skills.

Napier's rods

Napier was a seventeenth-century Scottish mathematician who devised a system of rods to facilitate multiplication. The system is shown in our diagram on p. 48, together with an example of a simple multiplication. Effectively it is similar to our adding the rows of numbers in a multiplication, except that we do not need to know our tables. Examine the rods themselves – they simply are the tables.

Napier's rods

	1	2	3	4	5	6	7	8	9
1	1	2	3	4	5	6	7	8	9
2	2	4	6	8	1/0	1/2	1/4	1/6	1/8
3	3	6	9	1/2	1/5	1/8	2/1	2/4	2/7
4	4	8	1/2	1/6	2/0	2/4	2/8	3/2	3/6
5	5	1/0	1/5	2/0	2/5	3/0	3/5	4/0	4/5
6	6	1/2	1/8	2/4	3/0	3/6	4/2	4/8	5/4
7	7	1/4	2/1	2/8	3/5	4/2	4/9	5/6	6/3
8	8	1/6	2/4	3/2	4/0	4/8	5/6	6/4	7/2
9	9	1/8	2/7	3/6	4/5	5/4	6/3	7/2	8/1

The rods: a way of writing the tables

To multiply 83 by 7, for example, take the two rods headed by 8 and 3. Look at the 7th row and put them together:

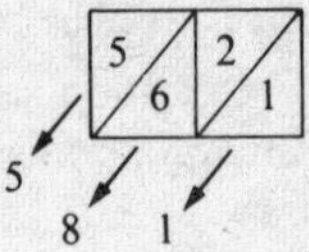

Answer: 581

Logarithms

The development of logarithms gave a tremendous boost to computation and opened up possibilities not available before. Again, much time has been spent on them in schools, though they are now obsolete as a method of computation, despite their mathematical interest. Again, it is the principle that is important, not the detail, and that is easily understood. They show another example of the importance of a decision to write something in a particular way. Mathematicians decided at one time to write numbers that were repeatedly multiplied in a certain way. In particular, with 10, the base of our number system, they decided to write

$$10 \times 10 \text{ or } 100 \text{ as } 10^2$$
$$10 \times 10 \times 10 \text{ or } 1000 \text{ as } 10^3$$
$$10 \times 10 \times 10 \times 10 \text{ or } 10000 \text{ as } 10^4$$

where the number at the top indicated how many tens were multiplied. On this basis we could write 10 as 10^1. We could also, and that is less easy, write 1 as 10^0, but we shall not need that for a basic understanding of logs.

If $10^1 = 10$ and $10^2 = 100$, what, we might wonder, is 83? Surely it must be 'ten to the one and something'! Using log. tables, or their successor the pocket calculator, we find

$$83 = 10^{1 \cdot 9191}$$

All numbers can be cast in this way as powers of ten. For instance, 236 will obviously be above 10^2 and below 10^3. In fact

$$236 = 10^{2 \cdot 2729}$$

Logarithms are therefore really a new way of writing numbers distinct from place value.

The important consequence is this:

$$10^2 \times 10^3 = 100 \times 1000 = 100\,000 = 10^5$$

so that we see multiplying *adds* the indices.

Returning to our two numbers

$$83 \times 236 = 10^{1 \cdot 9191} \times 10^{2 \cdot 2729} = 10^{4 \cdot 2920}$$

Looking in antilogarithms, or on our calculator, we translate this into 19588·447. Our earlier pencil and paper method gave us 19588 exactly. There is often a small error in working in logs.

The importance is that multiplication has been reduced to addition and quite heavy calculations can be accomplished. But with a calculator, why worry?

Calculating machines

Charles Babbage was credited with the first machine, 'a difference engine', to perform calculations, but technology in the first half of the nineteenth century was not up to the precision needed in the working parts. The basic principle is simply of gearing. If one wheel is made to turn once for every ten times of its neighbour, then we are already on the way to a physical model of place value.

Twentieth-century technology allowed the development of marketable machines, where we would frantically turn a handle, back and forth, and the numbers appeared before us. The machines were not wonderfully reliable, and were initially very heavy and clumsy. Later machines were easily portable but did stick if worked too hard.

Interest in these machines is now purely historical.

The slide rule

The slide rule was the invaluable companion of engineers and scientists throughout most of this century. Effectively it uses logarithms. When the 1 on the central sliding rod is placed against, say, 83, the numbers now read off on the upper scale are multiples of 83. This is achieved by making the scales logarithmic, that is equal gaps signify equal gaps in the logarithm of the number, not the number itself.

People accustomed to their use manage a great facility in dealing with complicated calculations, and some still retain them, despite the advance of pocket calculators.

The electronic calculator

Almost the ultimate weapon, the modern electronic calculator, that can be slipped into a small pocket, can do any calculation that the ordinary person might conceivably need. In a few years it has fallen from £100 or more in price to a few pounds, so that anyone who needs one can have one. Slightly more sophisticated models cope with the needs of the scientist, engineer or mathematician.

Even thirty years ago the idea that such facile calculation could be at everyone's beck and call would have seemed a fantasy. The resistance to their impact is typical of all forms of reaction. It was claimed that no one would in future be able to deal with numbers at all. In fact, the need is to move towards a situation where mental work and estimation is much extended, so that paper and pencil work is then totally replaced by the calculator. We hear nonsense about batteries wearing out (they do every couple of years or so) and the monstrous helplessness we would then face. It is all rubbish; the pocket calculator opens up new vistas of mathematics for all, clearing away the dull routine so that we may see the live, not the dead parts of the subject. In this book we shall need to caculate from time to time and shall turn to calculator and micro as naturally as to a pen.

The microcomputer

We have been talking of computation, but the computer does far more than compute. For the purposes of this chapter we shall take a brief look at what it can do that the calculator cannot.

There are parts of mathematics where a routine is repeatedly performed; it is known as an iterative process. We shall see in our chapter on algebra that some equations are impossible to solve in terms of a formula. Yet by finding (by guesswork) a result somewhere near our answer, we can devise a technique for getting a closer result. Using this closer result and the same technique we get one yet closer. The repetition is tedious even on a calculator, but a single loop on the computer allows us to progress with great rapidity to any desired degree of accuracy.

Another area is that of infinite series. Look at this string of fractions:

$$1+\tfrac{1}{2}+\tfrac{1}{3}+\tfrac{1}{4}+\tfrac{1}{5}\ldots$$

It is a series, and the presumption is that we take the obvious pattern of its development so that the nth term is $\frac{1}{n}$. Series like this can always be summed to a finite number of terms. If we set the computer up to add up terms we would take little time to find, say, 100 terms, and would probably get a million terms added up in several hours (depending on our micro).

Some such series obviously get bigger without bounds as we take more and more terms. A series such as

$$1+2+4+8+\ldots$$

is one that tends to infinity. As we add more terms we simply get bigger more quickly. It is surprising that the first series we wrote down also tends to infinity, even if it does so much more slowly. We can prove this, or we can watch what happens as we take more and more terms on the micro. Proof is more mathematical; seeing the numbers carries more belief.

In 1670 Gregory came up with the following result:

$$\pi = 4(1-\tfrac{1}{3}+\tfrac{1}{5}-\tfrac{1}{7}\ldots)$$

It is important that we establish π and indeed calculate it to as many places as we may need. Mathematically, the formula is interesting, easy to find (see Appendix, p. 231), and yet quite useless as a means of calculating. Until recently no one could possibly have got a remotely accurate value of π from adding terms of this series. Adding a million terms of this series took a microcomputer a few hours, and yielded the result

$$\pi = 3{\cdot}1415937$$

which is all right as far as the first five decimal places are concerned but not beyond. The task, however, taking only a few program lines, could not have been considered till now. Who can work out a million terms?

It may well be that we now have calculation wrapped up, and the story is complete, at least for nearly everyone. It does seem, however, that every new advance in mathematics, even in such a matter as calculation, can open new doors. Perhaps we shall need to reflect on these statements in ten years time.

4

Conversation with a computer

Our children are blasé about computers. They will grow up into a world where they are a standard adjunct to living. For the adult only just becoming aware of them there are mixed feelings, and as is common with new ideas, reactions are polarized and often lacking in balance. A few see computers as providing solutions to all our problems. More view them with considerable apprehension, and see a world where everything is 'taken over' by machines. Neither view is correct, although the eventual impact computers may make, particularly in education, is a matter for considerable speculation.

To play chess with a machine is a weird experience. A chess player is accustomed to sitting opposite another thinking being, a person, whom he sees as an enemy to be defeated – even if you have a drink with him afterwards. One cannot feel quite the same about a chess computer. At least it does not exude quiet satisfaction if it wins, but neither does it offer you a scotch. It certainly seems to think, and can produce the most surprising moves on occasion. This is perhaps our central concern. Does the machine really think?

What the machine can do is to hold a formalized expression of some *person's* thinking. That person may be very bright, and the programs that he or she devises for the machine complex, devious and difficult to combat. The machine is the medium for his ideas. We must not expect machines to think creatively, though we are on the verge of machines that genuinely learn from their own mistakes.

Properly to appreciate the power of, say, the microcomputer in the home, needs a long apprenticeship. Even in the simplest of programs, however, we can learn something of what such a machine can do. Let us converse with a computer.

We know, thanks to Eratosthenes, how to discover whether a number is prime or not. We divide it successively by 2, 3, 5, 7, 11 . . . and so on. If, when we divide, it goes exactly then the number, obviously, is composite. Furthermore, if we wish to test a number near 100, say 97, we only need to test as far as 10, because any larger number than that will leave as its other factor something smaller than 10. In practice, we only test to 7, the highest prime below 11. If we wanted now to test a number up near 10,000, we would only need to test using the primes up to 100 (which we know). In general, we go as far as the 'square root' of the number tested. That is the number multiplied by itself that would give the original number.

If we explained this to a person and asked them to test, say, 9371, then they would have a clear procedure, but it would take them a good while using pencil and paper methods for long division. Moreover, they might well make a mistake partway through and miss a factor. It is precisely these two human failings – slowness and unreliability – that the computer does not have.

Neither, however, does it have the flexibility of a person. Start a person on the procedure and they may find short cuts, ask pertinent questions, and know how to clarify their own difficulties. At this present stage of development a computer still needs to be told very precisely what to do, and in its own language. So let us discuss with it whether a particular number is prime. Language varies from computer to computer. Here we are talking to a Sinclair 'Spectrum'.

We first say to it

```
10 INPUT X
```

This means we type in 10 and then the command INPUT and the letter X. This prepares it to receive any number that we put in to be tested. All orders have to be numbered; it follows them in numerical order. We could simply put in 1, 2, 3, 4 . . . , but we often find there is an order that we should have put in earlier to make life simpler, and if we do not leave gaps in the numbering we cannot go back, without a complete re-numbering. So we normally number in gaps of 10 to give lots of room.

Now think back to our number 9371. If a person were working on it, we would tell them to start at 2 and work through up to its square root. The square root is just below 97, so in fact we do not need to go beyond 89, the next prime below that. If the computer has not got all the primes in its memory, let it test all the numbers up to 96. How do we tell it this? We say

```
20 FOR n = 2 TO INT(SQR X)
```

This looks more complicated than it is. SQRX is the square root of 9371 or about 96·8. The statement 'INT' means 'take the whole number part of' (or the INTEGRAL part of) and in the case of 96·8 this is 96. So we have given the computer exactly the same instruction as the person – except that we have courteously spoken to it in its own tongue. We have asked it to divide by 2, 3, 4 . . . 96 in turn to see if it goes exactly.

Our next task is to tell it to report if it goes exactly. This is where the 'INT' is again very useful. If we divide 9371 by, say, 37 we get the answer 253·27 (approx.). Had our input been 9435, again divided by 37, we would get 255 exactly. How can we ask it to distinguish?

We ask if

```
INT (X/n) = (X/n)
```

This is a nice device. If a number divides exactly the INT of the answer and the answer itself are the same. If it does not, the answer is bigger than the INT of the answer. Look at our two examples to confirm this. 253·27 is bigger than 253 while 255 is the same as 255. Our next statement to the computer therefore is

30 IF INT (X/n) = (X/n) THEN PRINT n; 'NO': STOP

Follow this through step by step. It is only when a number divides exactly that answer and INT (answer) are the same. If this happens for any n, we want to know that n, and wish to report that the original input is not prime. Also we can stop. The instruction given will mean that if the machine finds a divisor it will print it on the screen and write "NO" beside it.

The 'IF' is a strictly logical statement, and we shall look at logic in a later chapter. The machine reads it, and if the condition (i.e. INT (X/n) = (X/n)) does not obtain, it steps to the next instruction.

40 NEXT n

This is the same as asking a person to try the next number if that one does not go. It will then 'loop' back to try the process for the next number, until it gets, in this case, to 96. When it does so, and only then, it will move on from instruction 40 and needs to be told what to do.

50 PRINT 'Yes'

It has tried all the numbers; they do not go, so the number is prime. End of question.

This is a very short program, clumsy in the method it uses to determine whether a number is prime, but taking only a few seconds to check on whether 9371 is prime. In fact a number like 987,353 rates only about half a minute to check. So fast is it that we need hardly bother that the program is clumsy.

The process is speeded greatly if we get the computer to find all the primes up to, say, 1000 by a clumsy method, record them, and use only them, not every number, in testing above 1000. This is not difficult to program. The aim here is not to teach programming but to link computers with our various other activities in mathematics, so let us simply say that as we find new primes they can go into store to test higher still. Knowing the primes up to 1000 allows us to test all the numbers up to 1,000,000 because a thousand is the square root of a million. Eratosthenes lives again!

As far as numbers go, the machine is quicker and much more reliable than you are. Yet it only 'thinks' according to a routine laid down by a human being. As with the chess-playing machine, it has in it things expressed by a person, and often one who is very clever. If we do not understand his thought processes, we may not understand how the machine does the operation. It is hoped that the instructions we gave about finding primes show us the principle. Finding primes is not that difficult, but many people may not know how to do it, and in consequence may be mystified by the computer, which is in fact merely doing fast calculations on rules laid down from outside.

We have mentioned from time to time the value of the pocket calculator. It is worth at this stage comparing its power with that of a computer on the problem we have just looked at. Take the simple calculator, that we may now buy for a few pounds. Finding if 9371 is prime involves much more on the part of the user. He does not need to know how to program, but he takes rather longer. Knowing that he only needs primes up to 89 (the one below 96) and knowing what they are from the table in our chapter on the 'basics', we set about it thus. We enter 9371, put it into memory, divide it by 3 and see if the result is a whole number. If not, bring it up from memory again, divide it by 7 (we know 5 is no good) and so on. There are only 22 to try (excluding 2 and 5) and it only takes a minute or so.

Yet it does not compare with the micro. Once the program is in and available, all we do is plug in a number and it very rapidly says yes or no. It will not balk at large numbers, and perhaps most importantly, if it has a lot of work it will do it without annoying you, while you get on with something else.

There are intermediate stages between calculators and micros, with programmable calculators, but they are a hybrid, likely to die out. If you need to program, you need a micro.

The micro can calculate like a dream, but that is only one of its parlour tricks. Its ability to sort and order is incredible. We shall see in our chapter on 'Modern Mathematics' how important is the ability to classify. Put in six hundred names in a subscription list, suitably labelled, and a computer can put them in alphabetical order, sort them into men and women, young and old, or where they live, in two shakes of a lamb's tail. Moreover, the size of the task, provided it is within the memory capacity of the machine, does not worry us. Ask the machine to get on with it and come back when it has finished; rather like a washing machine. Household accounts, one's private library are all meat and drink to it.

Most of us will not want to program; the description of a program to find the primes was to illustrate the way we need to address our micro. The routines we have just been describing are indeed programs, but we do not need to write them. The 'software' experts have written them so that anyone can handle the machine. Once the machine is equipped with programs it asks you questions, tells you how to type in answers and generally treats you as if you, and not it, are an idiot.

We certainly do not want to get into technical jargon, but we have used the word 'software' and we should not introduce words without explanation. The actual box, metal or plastic, and the attached wires and TV screen are all 'hardware'. The instructions, written on a disc or on a tape, are the 'software'.

Many people buy a micro, and various accessories, like a printer or a disc drive, or some extra memory, and have some very efficient hardware. If they are not computer buffs interested in the mechanics and programming, all they may have is a few games supplied by the computer company. Some of these are fun. They require quick reflexes and great concentration, but they are ultimately a mindless occupation. For most people the novelty wears thin after a time and begins to pall. Soon after that the computer and its accessories appear in *Exchange and Mart*. The need is for good software.

We have indicated the use in filing systems, but there is a much extended capability which we shall all shortly explore. Think of some of the following possibilities.

Firstly, information retrieval, to give its rather up-market name. If you want the whole of an encyclopaedia on disc so that you can look up anything at any time, the computer can handle it. We may not be fully organized yet, but we are certainly beyond the stage of some early material that asked us what is the capital of France and congratulated us when we chose Paris out of a shortlist of four.

Secondly, there could be technical manuals, with the growth of graphics on our screens. It will not be long before a program could be prepared that told you stage by stage how to change the clutch on your car. It is, of course, in a manual already, but the movement and life that can be instilled into a well-devised computer program puts it far ahead of written material. Not least is the ability to suggest three-dimensionality.

We already have extensive systems of word processing, which allows us to shift chunks of prose from place to place, deleting this and adding that, and printing out a very neat piece of work at the end. Bang goes the typewriter.

This can be extended into spatial and geometric areas. When you are sorting out your new kitchen, why not do it on the screen, moving things about to see how they would look from different angles? Get the machine to tell you how you can best cut an awkward-shaped carpet moved from one room to fit another. Set up a picture of the walls of your room and it will tell you how to wallpaper it.

Having said all this, and having perhaps touched on an educational use in information retrieval, let us state quite boldly that the impact on education may be such as to transform our institutions and our social fabric. Great claims have been made before of the dramatic effects of various technological advances on the way society operates. Some effects have been less than forecast but some rather more. Did even Henry Ford foresee the pluses and minuses of the impact of the car?

Let us start by looking at how earlier advances have affected our educational system. Radio and television were expected to revolutionize teaching and learning. They yet may. The move to disc video material is beginning and we may soon be able to view at will any of the range of material now available on the various channels. Till now they have been confined not only by cost but by broadcast time. The constraints of this are horrifying. The presenter is told: 'All the visual you like, materials and resources, but just sort out negative numbers for everyone in fifteen minutes.' A challenge, but a bit demanding. Before long we may be able to make much longer programs, no longer broadcast, but available in some form of recording for our home.

The computer software experts are not going to match that visual impact, but they have one very important educational advantage: the programs are genuinely interactive and involve the student. The computer can ask questions and make demands that no film or video can do. When we were discussing factors in 'Back to Basics' we showed a game of some difficulty based on the tables. The program for this is moderately intricate, but once the machine is set up, it plays the game accurately and with great courtesy.

We are still at the beginning as far as software is concerned, and we do not know where it may go. There are scenarios we may imagine. With the advent of cable and the possible development of banks of central material we can envisage in a few years the capability of sitting at home and calling up a program on anything at will. If we decide that we need to brush up on quadratic equations, programs that involve the participant in games and strategies aimed at understanding and applying them will be available.

A significant feature of this facility is that individual students are in control of their learning. They decide on the pace, the content, the level at which they can cope. It is not suggested that this will result in the whole population suddenly becoming students, but it opens doors for many who have found the methods of our schools unhelpful. That is not to say that the schools do not try, and with some success, but not all people are adapted to the processes to which we submit them. No institutional system can cope with all types of individual. We should therefore welcome the addition of well-planned computer programs to the system because they become accessible to everyone at home.

II

The central core

5

Number

Number is at the same time the most familiar and one of the most difficult of areas. By looking at it we shall be able to see with great clarity how mathematics expands. The extensions that we make as we dig further and further into number reveal a process that is repeated in many other mathematical topics. We have already discussed some issues in number. We have looked at the tables, the primes, counting, various forms of numeration and place value. In doing so we have taken some matters for granted that warrant another look. We must proceed in easy stages.

The natural numbers

These are the counting numbers; in our way of writing them, 1, 2, 3, 4, 5 . . . . We know that counting has more to it than we might have thought, and that the way we decide to record numbers has great significance. Here, however, though we have to display and discuss using the symbols, we are not interested in the numerals, but in the numbers themselves. We are interested in the number 7, whether written as we have just done, or as 'seven' 𝍸II, VII, 7 or 'sept'. Underlying the ways of writing it is the thing itself, an abstract number, prime or whatever, no matter how we write it.

There are two ways of looking at how number came to be. The simpler, historical one is that it arose from a social need, as we have already discussed. It was simply a device for keeping track of whether we had got everything. It hinged upon possession; we had to own things before we needed to check whether we had them all. In this view we presumably invented it, in a way similar to that in which we invented other tools.

Another view is that number is fundamental in the world we live in, that it is objective knowledge which any race sufficiently intelligent is bound to develop. In this sense it is discovered rather than invented. We shall return to this discussion much later in our last chapter. The distinction is one we need simply to note at this stage.

We now, as it were, cross-reference number to a spatial idea, that of a line with equally spaced points – the number line. We seem to have done nothing special; this representation is familiar to us all. Yet it was conceived at some time; the link was invented. We shall find this image of number a potent one; images raise questions, and the questions vary according to the image we use.

We now have two quite separate images. The first is of a collection of objects – and we are not concerned as to what the objects are – and the second a line with a series of marks upon it. Each is a representation of number, neither is number itself. Number is essentially abstract, yet it is something we have come to terms with even before we go to school.

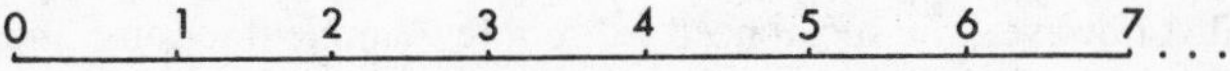

The number line

At some stage, and this hinges upon our learning the system of numeration, we begin to realize that not only are numbers abstract but they go on for ever. There is a sense of awe about this; the notion of never stopping is difficult to accommodate. If our image is a number line, we are soon in speculations as to whether space goes on for ever. The human race has wrestled with the notion of infinity and come to some conclusions, but it always remains disturbing. There is excitement too. Most children go through a stage of looking for larger and larger number names – billions and trillions and so on. A spirit of competition enters the stating of large numbers, but we soon find that no one can win.

There are many mysteries within the natural numbers, particularly associated with the primes, but we have grown accustomed to them, and feel some sense of ease. This is enhanced by certain numbers becoming personal to us. Our age, or perhaps our shoe-size. Constant usage makes us feel moderately friendly to the positive whole numbers.

Fractions

For many purposes natural numbers suffice. Many real life objects come in whole numbers, and need no other numbers. Yet an extension became necessary, and it led us historically to fractions. There are various ways of looking at fractions; here we shall examine three main thrusts that demanded they be developed. Each thrust looks quite different at first, but each leads to the same eventual structure. The three are: social need, the desire to get answers to operations, and geometrical analogy.

Socially we have been obliged to share things. The most commonplace sharing is to break something into halves. A 'whole' (or the number) is being rendered into smaller parts. The beginnings of social organization must lead to this idea, and to the development of language to say what is being done. It is not easy to conceive of a language that does not have a word for 'half'. Sharing a whole between a number of people gives special significance to certain fractions: those with one at the top, like $\frac{1}{2}$, $\frac{1}{3}$, $\frac{1}{5}$, $\frac{1}{11}$ and so on. The Egyptians, in their calculations, paid special attention to such fractions. We call them the 'reciprocals' of the counting numbers, and we usually find a key on our pocket calculator that does just reciprocals for us (that is, it works them out as decimals, and more of that anon). We can devise rather artificial social situations where 2 or 3 wholes are divided among more people than that. Generally, however, the social need led to a few simple fractions, and that is true even today. We need to share, we need the language of halves, quarters, thirds, tenths, but we need to go no further in our practical everyday living. The first impulse to fractions, then, is social.

The second is related; naturally it has to be if it is to lead to the same place. It is, however, a little more sophisticated. Once we had got hold of the counting numbers we began to calculate with them. We soon had the numbers and also 'four rules' to allow them to operate with each other. It is useful at this stage to separate the notion of the number and the operation clearly in our minds, and to look at each separately. The operations and the symbols we use to represent them have, as it were, a life of their own. They can be studied as entities separately from the numbers on which they operate. This will become clearer as we penetrate later into algebra.

If we add two natural numbers, the result is a natural number. Never be afraid to state the obvious. In mathematics we often start with such a statement and within a line or two are in deep water. If we multiply two natural numbers, the result is always a natural number. A system where an operation on two elements produces another element of the sytem is called 'closed'. Our second obvious statement now forces us to think.

At the moment we do not want to look at subtraction; we shall merely remark that this time we do not get closure. Sometimes we get a counting number by subtraction, sometimes not. And at the moment counting numbers are all we have.

Now for division. Division we know is related to sharing, though we are now not thinking of a real world, with food to share, but of the abstract world of number, where certain processes go on which may or may not have connections in the real world. Simply try dividing one natural number by another. If you are lucky, you get a natural number ($12 \div 3 = 4$), but mostly we do not ($5 \div 7 = ?$). It is no good saying we get a fraction; we have not invented them yet. If we are restricted to the natural numbers, there is no answer to $5 \div 7$. It 'doesn't go'. Children will tell us this very forcibly if they have not been introduced to fractions.

For those of us accustomed to fractions, at first it seems a quibble. If that is your feeling so be it, but you may need to return to this passage later if you find complex numbers a block.

We meet the need for an answer by 'inventing' fractions. We say that $5 \div 7$ is $\frac{5}{7}$. Ask many a person what that horizontal line means and they will say it means 'divide'. So in what sense have we invented any new numbers? We have simply ducked the question by writing it in a different way and pretending we have some new numbers. Indeed, that is how it seems. Our next step is to see how these new numbers operate. We discover that, for instance, $\frac{3}{6}$ is the same as $\frac{1}{2}$ and 'cancelling' enters the scene. We find that we need some very odd-looking rules. For instance

$$\frac{2}{3}+\frac{5}{7}=\frac{(2\times 7)+(5\times 3)}{3\times 7}$$

What a crazy idea! It is no wonder that we are all bemused when we are first asked to add fractions. There are many ways of making it seem easier and we deal with adding fractions (the hardest operation) in the Appendix (pp. 231–2).

The interesting thing about the rules we make is that they have to be internally consistent. We do not need to go to the real world, we just need to make sure we do not get contradictions. These fractions come out of the whole numbers and the operation 'divide'. The essential thing is that the rules must still apply to those whole numbers, and for that we need a new way of looking at them. We therefore regard them as fractions, but with 1 at the bottom. So 2 is $\frac{2}{1}$ and 7 is $\frac{7}{1}$. We do not *have* to do this, but it brings them into line and gives a helpful new image. The natural numbers are not now seen as separate from the fractions, but as a part of them, a special enclosure inside the main enclosure into which we put all the fractions with denominator (bottom line) unity.

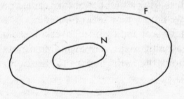

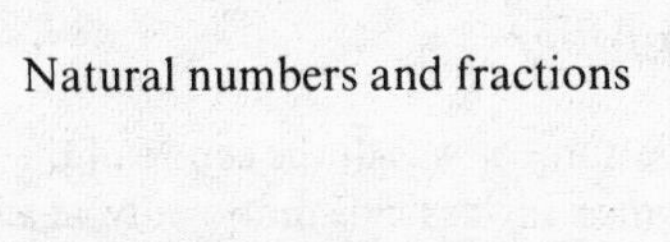

Natural numbers and fractions

If students spent their time understanding these connections between natural numbers and fractions, and how the need arose mathematically for fractions, they would be better equipped to understand mathematics than if they could work out

$$\frac{\frac{2}{3}+\frac{7}{11}}{\frac{5}{19}+\frac{16}{17}} \quad \text{(which is no use to anyone)}$$

The third impulse is less directly aimed at fractions, but provides some interesting new insights. Once we had drawn a continuous line on which to put the natural numbers, we raised questions about other points on the line. If it is a railway track with stations, well and good; the dots are then a different type of thing from the rest of the track (although the stations still need the same track). Do the points between the counting numbers represent numbers? The answer is 'If we so choose'. We are deciding what numbers are, and if the line represents numbers, then we shall look further into that line to find out more about them.

An obvious question is, 'How close together are they?' The answer seems almost as obvious as the question; they must be as tightly packed as we like. If we take two of them, say $\frac{2}{3}$ and $\frac{5}{7}$, which may not be all that close, we can certainly find one partway between them, in this manner:

$$\tfrac{2}{3} = \tfrac{14}{21} \quad \text{and} \quad \tfrac{5}{7} = \tfrac{15}{21}$$

so we want something between 14 and 15 twentyfirsts. That is easy enough. We take $\frac{14\frac{1}{2}}{21}$ and simplify to $\frac{29}{42}$. We now have $\frac{2}{3}$ as $\frac{28}{42}$ and $\frac{5}{7}$ as $\frac{30}{42}$, leaving a nice simple centre point at $\frac{29}{42}$. This we can always do. What is more we can find something halfway between $\frac{28}{42}$ and $\frac{29}{42}$ in a similar way, and continue narrowing the gap as much as we like.

We say that fractions are 'everywhere dense' on the line.

At this stage we have broken our original bounds, the set of natural, counting numbers, and arrived at some sort of rationale (no pun intended) for the fractions. We have subsumed the natural numbers in this larger entity, and pretended they too are fractions. We have found that there are so many of them that our line is beginning to look crowded. To return to social need, we could now deal with sharing a number of cakes among a lot of people. Using all three impulses, we have a new sort of number.

If we seem to have been indulging in 'overkill' and making much of little, wait a while. The reasoning so far may seem simple, but there are traps ahead. The first of these shook the Pythagoreans to the roots of their beliefs.

Algebraic numbers

For the Pythagoreans, number was at the centre of the universe; this was not only an intellectual stance, it was held emotionally, at a mystical and religious level. Eventually everything had to derive from the natural numbers. Around the natural numbers were woven many theories. Numbers were invested with character and power to affect the world. They were lucky or unlucky, or 'perfect'.

A perfect number is one whose factors (including 1 for reasons not very clear) add up to the number. The lowest of these perfect numbers is 6 (for $3+2+1=6$). The next is less than 100, and the next rather big – but if anyone wishes to search, it is quite an interesting occupation. If you find it tedious, use a home microcomputer; if you instruct it properly, it will search a good deal quicker than you will.

Their whole attitude was of a priesthood, who guarded the truth. The truth was that everything was based on natural numbers. They accepted fractions readily enough; after all, every fraction was a ratio of two of these basic elements, and it was assumed that this was all as far as number was concerned. Then they looked at the diagonal of a square with side 1. They knew (and it is Pythagoras' theorem) that the length of the diagonal was $\sqrt{2}$. In our diagram we have drawn the unit square, and its diagonals. Using a compass we now mark on the number line the distance $\sqrt{2}$. Again we may be stressing a point, but $\sqrt{2}$ is on the number line.

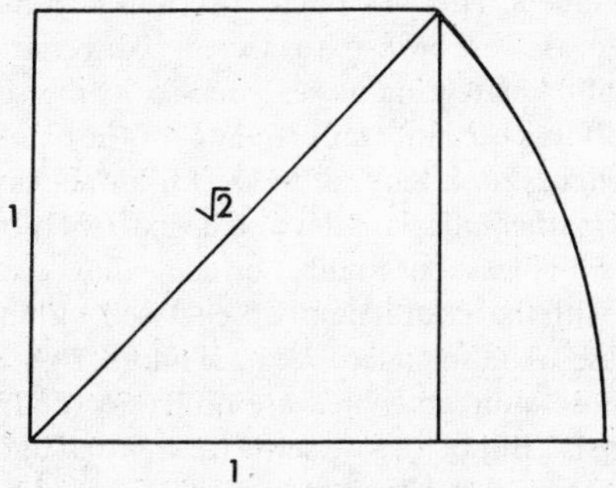

Pythagoras' theorem

The discovery that shattered them was that $\sqrt{2}$ was not a fraction. That they discovered it is extremely impressive; it is one of those examples in mathematics where a proof is short, elegant, easy to understand, and yet required a genius to discover it in the first place. It rivals in simplicity Euclid's proof that the primes went on for ever. It is worth stopping to see if you can do it before you look at the next few lines.

They said, suppose

$$\sqrt{2} = \frac{a}{b}$$

where $\frac{a}{b}$ is a fraction, written in its lowest terms (that is, cancelled down; we would not accept $\frac{15}{27}$ but reduce it to $\frac{5}{9}$).
Square and we get

$$2 = \frac{a^2}{b^2}$$

or

$$2b^2 = a^2$$

They now reasoned in this way. The left hand side is even, so the other side must be. The only way a square number can be even is if the number itself is even. So put 2 = 2n where n is a whole number.

$$a^2 = 4n^2, \text{ so}$$
$$2b^2 = 4n^2$$
$$\text{or } b^2 = 2n^2$$

Now we reason back as before. The right hand side is even. So then is b. We now have a and b both even – and this is a contradiction since the fraction would then cancel down. So $\sqrt{2}$ is not a fraction.

Notice the style of the reasoning. It is very similar to that with the primes. We accept the opposite, find a contradiction, and are led to a positive proof.

There were numbers that were not fractions, and did not therefore come from the whole numbers. This was a matter serious beyond belief. The whole school of Pythagoras were sworn to secrecy concerning the dread fact. Fortunately or unfortunately, however you see it, the secret was revealed outside the circle. It was certainly unfortunate for the character who did leak the news; his colleagues took him out to sea for a one-way trip.

It is a serious mathematical matter and an extremely interesting one. We have to find room for at least one number on a line that is very densely packed. Also our diagram with the natural numbers enclosed within a larger enclosure of fractions will have to be amended. We need another yet larger enclosure. It will contain all those numbers which are not ratios or fractions. They are called the 'irrationals'. But before we settle for a broad and rather negatively defined category, take a closer look at numbers like $\sqrt{2}$.

In our algebra text books (they seem to be put in algebra rather than arithmetic) they are often called 'surds'. Following the same policy as with fractions, we could say $\sqrt{2}$ does not 'go' because it does not come out to be a fraction. If fractions are all we have, there is no answer. As before, we write what we intend to do, $\sqrt{2}$, $\sqrt[3]{3}$ or $\sqrt[5]{71}$, but do not actually 'work it out'. We then work with the expressions, and see how they operate. For instance

$$\sqrt{2} \times \sqrt{3} = \sqrt{6}$$
$$\sqrt{2} \times \sqrt{8} = \sqrt{16} = 4 \text{ and so on.}$$

This is exactly parallel to the steps we took with fractions. We said it did not go, but assumed there was such a number, then examined its behaviour. Our reiteration of this policy has purpose. Numbers which consist of any combination of roots, of any power, are called 'algebraic numbers'* because they can all come from equations (see Chapter 7 on algebra) with whole numbers as coefficients. Maybe the Pythagoreans would be cheered that whole numbers again seem to be important. We now see $\sqrt{2}$ as the answer (or one answer) to the equation

$$x^2 - 2 = 0$$

* We shall consider only 'real' algebraic numbers.

Negative number

We now need to track back a little. Historically, the issue of $\sqrt{2}$ was a good starting point, but we have run forward from the Pythagoreans. Equations have been a fruitful source for advances in mathematics, and as early as the third century AD Diophantus (mainly known for giving his name to equations involving just whole numbers) came across -4 as the solution of an equation and rejected it as absurd. This attitude persisted, even among well-known mathematicians, as late as the middle of the sixteenth century, when Cardan recognized, 'minus times minus gives plus', but still regarded negative numbers as 'fictitious'.

We can start with an equation. Toss a weight to fall three feet away. It starts off, say, 6 ft up (we still measure in Imperial) and the flight looks like that on the graph below. It is *part* of the graph

$$y = 6+x-x^2$$

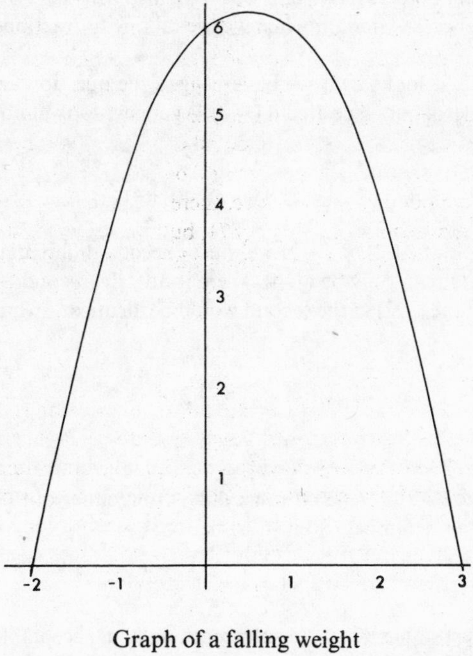

Graph of a falling weight

The graph is mathematical (we may need to refer forward to Chapter 7 on algebra to solve quadratic equations) but the situation is in the real world.

Where they overlap, they fit. That part of the graph starting from 6, going slightly up and then coming down to hit the ground at 3, accurately represents the path of the weight. The mathematics is unaware of the real life situation, and says that the graph (a parabola) hits the ground also two feet behind you. It makes mathematical sense, for the mathematics extrapolates in both directions. It is a fact that the expression which exactly fits the flight happens to be zero at 3, an answer we are interested in, and at -2. The second answer does not fit the real situation in this case – but it is still good mathematics.

Ignoring their patent absurdity, we shall make our usual three-pronged attack on the negatives. This time we shall start with the formalist approach. We are back where we only know the counting numbers, and we have been using the four rules on them. We were not happy about division and it led us a long way. Neither were we happy with subtraction. Sometimes it worked and sometimes it did not. So $3-5$ has no answer. We adopt the policy we had with fractions. We invent a new number, called a negative number. It might have been simpler if we had called it $\bar{2}$ or some other symbol. In fact, because mathematicians have a taste for confusing things, it was called (-2). This mixes up the minus sign meaning 'subtraction', an operation, and the number itself. The sign of a negative number is built in, it is not an operation. To emphasize that numbers have signs we might also write $(+2)$ for our ordinary counting number 2. After all, when we are talking of fractions we call 2 by a new title $\frac{2}{1}$.

As before, it looks as if we have begged the question and not actually taken 2 from zero, but just called it (-2). Again, we look at how it works. We are happy that

$$(+2)\times(+2) = (+4)$$

Now if we look at $(+2) \times (-2)$ we must expect a different answer. It must have size 4, so it can only be (-4). Were it not, the second $(+2)$ in the first statement and the (-2) in the second would be identical – which they are not. So

$$(+2) \times (-2) = -4$$

and also

$$(-2) \times (+2) = -4$$

If now we take $(-2) \times (-2)$ it must differ from either of these, because we have changed one number. So it must be that

$$(-2) \times (-2) = +4$$

The logic is irrefutable but the result emotionally unacceptable. Re-read that sentence, for it is one of the things that makes mathematics hard to come to terms with. In all results, we need to believe that they make sense, not just that they can be proved. It is a problem. Mathematics does not need to be justified by a real world situation, but let us look at some, and see if it helps belief.

In our modern world we have become more accustomed to negative numbers. We are fully aware that an overdraft at the bank can be thought of as a negative number, as can a temperature below zero. But there is no real-life meaning to debt multiplied by very cold weather. It is the operations between negative numbers that we need to understand, and there are some models.

The diagram on p. 70 shows the sliding price of a car. It is a bit artificial, since it would not usually be as steady, but it is not unreasonable. The present moment is zero and we paid £1200. The car will depreciate at £200 per year. It is surely natural to call that (−£200) per year. An appreciation (as with house prices) would be a positive number. A depreciation is a negative appreciation.

Time always has a direction. The future is + and the past −. That accords with our general beliefs. If we want to find how much the value differs from the present price we multiply time by appreciation. If we look to three years hence (+3), we get

$$(+3)\,(-200) = (-600)$$

and the price is £600 down – which agrees with common sense.
If we look at the situation 3 years ago (−3) years the difference is

$$(-3)\,(-200) = (+600)$$

and this tells us, as we know already, that it was worth £600 more. The tag (−£200) is permanently attached to the car. We can look backwards or forwards in time, but looking backwards 'minus times minus gives a plus' makes sense.

Latching on to one single case like this can make the whole notion acceptable. An alternative parallel from a different source is the double negative in language. To say

'There is no way I am not going to be there'

may be clumsy, though perhaps said for emphasis. The meaning is

'I am going to be there'.

The two negatives give a positive. Different parallels appeal to different people.

The thermometer image is a useful one. It connects with the number line (drawn vertically rather than horizontally). The line itself raises the question: why do we start at zero, and could we not travel back in the other direction? As before, social, mathematical and geometric analogy all lead us to negative numbers.

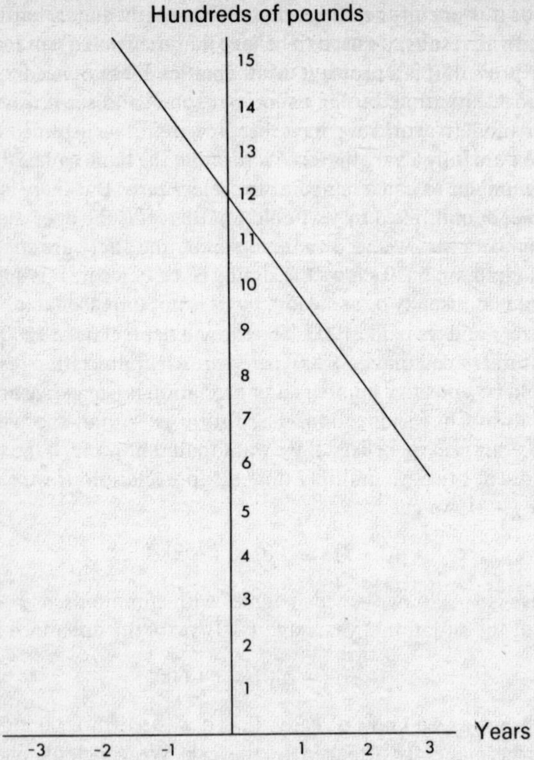

Car depreciation

Transcendental numbers

We are building up a formidable categorization of number, and we now briefly look at some of the most intricate. It starts with one of the most simple of shapes, the circle. If, as in the diagram, we have a circular pond, we are well aware that it is a lot further round the edge from A to B than straight across. In fact, more than half as far again. To go right round is over three times as much as one trip across. This has been known for thousands of years. In primitive times this ratio was taken as 3, which was good enough, even if it was recognized it was in fact a bit more. The Greeks gave a name (one of their letters) to this proportion. They called it pi, written π.

There is an issue, however. Is it fixed, or does the proportion depend on how big the circle is? Suppose the circle were the Round Pond in Kensington

Gardens, or Stonehenge. Now look at a plan or photograph of either place. Surely we expect the proportions in a map to stay fixed? Then the proportion of circumference to diameter stays fixed, irrespective of size of circle. π is fixed. It would be rather convenient if π had happened to be a whole number, or at least a fraction. Unfortunately we have no control over what it is. It is what it is – and that is a rather awkward number. There are fractions that are tolerably good approximations to π. The best known is 22/7, which is not all that close. The Chinese knew that 355/113 was more accurate – in fact remarkably accurate. A computer program to find good fractional approximations to π is given in the Appendix (p. 232). But the point is that π is not a fraction.

We are accustomed to that view; we know that some numbers are irrational. Algebraic numbers come as the solutions of equations with whole

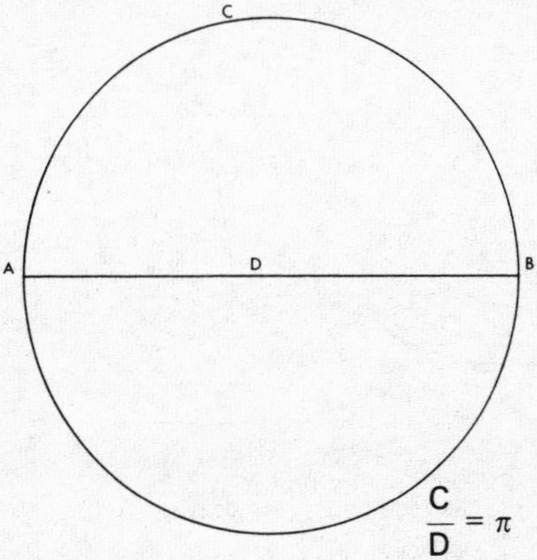

The circle and π $\frac{C}{D} = \pi$

number coefficients, but π cannot even be got at that way. The Greeks successfully drew squares equal in area to any triangle and hoped to do the same with the circle. 'Squaring the circle' has become a well known phrase. In another of the interesting examples where mathematicans placed deliberate constraints upon themselves, they asked that only straight edge and compass be used. Certainly this permits at least some algebraic numbers – such as $\sqrt{2}$, obtained by drawing the diagonal of a unit square. In 1882 Lindeman proved that π was one of a new class of numbers—'transcendentals' which did not come from an algebraic equation. The other well-known member of this category is e, which we meet in the chapter on the calculus.

So irrationals consist of the algebraic and transcendental numbers. It would be wrong to suppose that because we have mentioned only a few of them, therefore they are rare birds. In fact, and in so far as we can make such remarks about infinite collections (and we can), there are vastly more irrational numbers than rational. This offends our geometric sense, because the fractions were packed so closely on the line, yet it is so.

However, the reader may be relieved to know that we have now filled up the line. We now have what are called the 'real' numbers. The smallest category is the counting numbers. It is easiest to make the first extension to negative whole numbers even if this does not follow historical development. Next we subsume all the whole numbers as a subset of fractions. We tend to define the irrationals in a negative sense – all those that are not fractions and split them into transendentals and algebraic. Our diagram sets it out. We assume all numbers can be positive or negative. Algebraic numbers, deriving from equations, can in fact be fractions and whole numbers. Transcendentals are defined as not being algebraic. It would be neater if our naming were different, so that our last category was more extensive, including all others. We do call the whole lot the 'real' numbers.

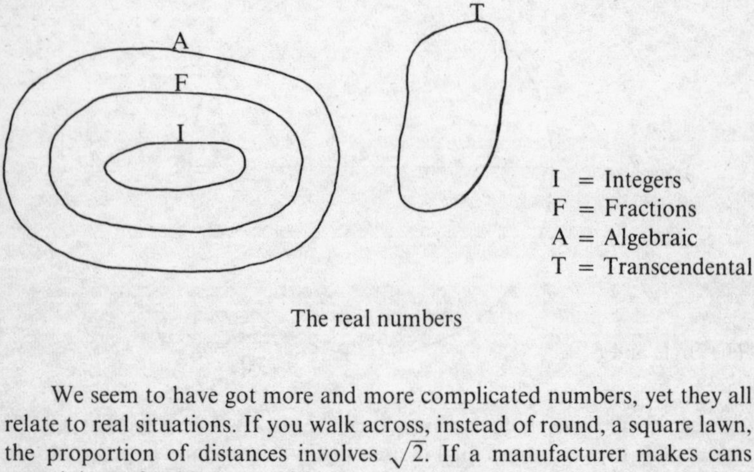

The real numbers

We seem to have got more and more complicated numbers, yet they all relate to real situations. If you walk across, instead of round, a square lawn, the proportion of distances involves $\sqrt{2}$. If a manufacturer makes cans containing a fixed amount of food or beer, he needs π to work out the best shape to minimize the metal he uses. It may not matter to him that π is transcendental, but it is fun for the mathematician.

Decimals

Have we missed something? What are decimals? In a sense, they are not new numbers, but ways of writing numbers. It was something of a historical deviation that we even got into fractions; we developed this in our chapter on calculation. Yet it is certainly worth asking how the numbers we have looked at appear as decimals. Integers (positive and negative whole numbers) are in decimal notation. Fractions always appear as terminating or recurring decimals. For instance

$1/16 = 0{\cdot}0625$ which stops

$5/7 = 0{\cdot}714285714285714285\ldots$ an extensive stutter.

All fractions appear either as decimals which finish, or get into a repetitive form, however long.

It is characteristic of irrational numbers that when they are expressed as a decimal there is no way of determining how the expansion goes. There is no recurrent pattern. In particular, transcendental numbers like e and π have been calculated hundreds of decimal places, with no apparent pattern to the way successive numbers appear.

Looking from the standpoint of decimal representation, the expansions without pattern are transcendental and algebraic, those with pattern are fractional, and those which terminate are a special sort of fraction, whose denominators have factors only 2 and 5, however many times each is repeated.

The square root of minus one

This is seen by many as the ultimate mystery. Is there some sort of initiation ceremony for mathematicians, involving their meeting this weird object? It is all nonsense, there is no mystery; it is no more difficult than any other number, and is of practical value in the real world. Let us meet this mysterious entity.

Our wearisome repetition of the 'formalist' approach now yields fruit. When we could not divide 5 by 7 we invented fractions – they had a social meaning but it did not matter mathematically that they had. When we could not do $5-7$ we invented negative numbers. We did not work them out, we simply called them negative numbers, and worked with them. When $\sqrt{2}$ did not 'go' in fractions we invented the algebraic numbers. We studied how surds worked, rather than worrying about their existence.

Back to our equations. Take this simple one:

$$x^2+1 = 0$$

Draw the graph on p. 74. We find roots when the graph crosses the x axis; it does not cross, as we see in our diagram. We make another start. Treat the operation of square root in the way we dealt with divide and subtract when they 'did not work'. As before, duck the question.

$$\text{Let } \sqrt{-1} = i$$

In a sense that says nothing, but we make it meaningful by working with it. All we need is that $i^2 = -1$, so let us look at some results.

$3i+4i = 7i$ Why not?

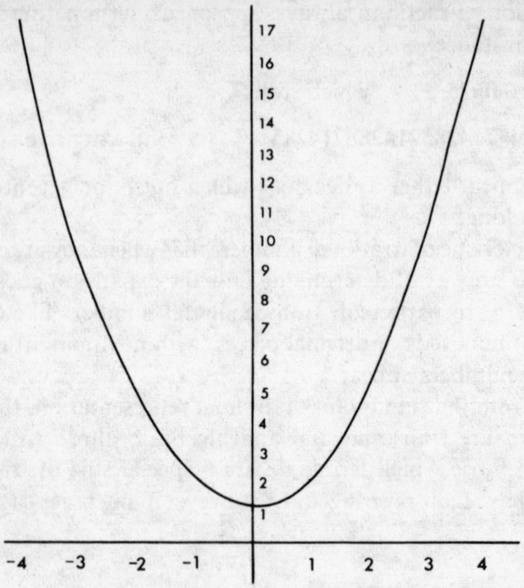

$$y = x^2 + 1$$

$4 + 2i$ cannot be worked out; leave it as it is. Now multiply it by i
$4i + 2i^2 = 4i - 2$.

So far so good. We have put $i^2 = -1$. It is no different in principle from what we did with fractions. We avoided the issue and saw how the new number worked.

It is intellectually compelling but emotionally unacceptable to adopt this approach, despite the fact that we have adopted exactly the same policy in earlier extensions. There will be a demand that we get a geometrical analogy (even if we have filled the number line), and get some social relevance. Mathematicians cannot guarantee to meet the demands in general; it is possible here. Our diagram on p. 75 shows the geometry. We put the 'imaginary' numbers on the y axis. A 'complex' number is made up of a 'real' part and an 'imaginary' part. $4 + 2i$, which we said above cannot be worked out, can be thought of as the point (4, 2) where x is the real part and y the imaginary part.

We now attach a geometrical meaning to adding complex numbers. Take a second number $(2 + 3i)$. We simply add

$$4 + 2i + 2 + 3i = 6 + 5i$$

The rule is that we add the real parts and the imaginary parts separately. Geometrically we get a parallelogram – which fits exactly the way that certain physical things like forces add together.

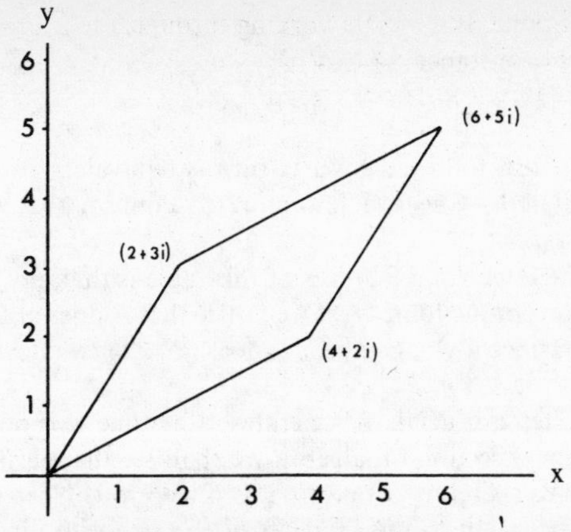

Geometry of complex numbers

Adding displacements, adding forces, all can be interpreted using complex numbers. What is even more significant is the fact that electronic engineers could not operate circuit theory without a constant use of $\sqrt{-1}$, although they always call it j. Much of modern technology uses it without any question, or doubt, that it is relevant.

Being told that this is so is still unconvincing, so let us try for one last everyday model to make these numbers acceptable.

You are travelling through a city built on a grid system. Your passenger is directing you. At every junction he offers one of four commands, L, R, U, A, meaning 'turn left', 'turn right', 'do a U-turn' and 'go ahead'. Examine the effects of successive commands on the direction you are going. How does

$$LL = L^2 = U \text{ match your perception?}$$

It means 'If you turn left and then left again you are going in the same direction as if you had done a U-turn'. True enough.

Similarly

$$R^2 = U$$

We seem to be into algebra, rather ahead of time, but we hope that the interpretation is clear. We can now set up a table for these commands:

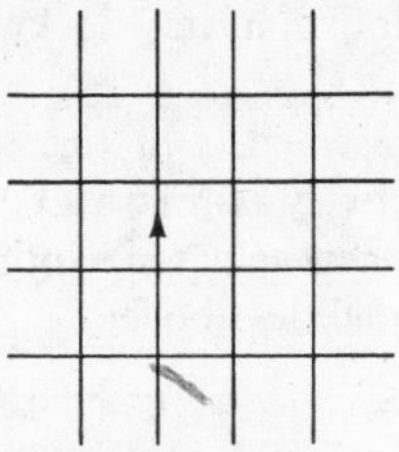

City streets

	A	L	U	R
A	A	L	U	R
L	L	U	R	A
U	U	R	A	L
R	R	A	L	U

An interesting table. It is worth working through a few 'products' to see if it makes sense. For instance

$$RU = L$$

means a right turn followed by a U-turn is equivalent (for direction) to a left turn. And it is. Check a few more to confirm that the table makes sense.

One of the interesting features of this table is that any two 'multiplied' together give one of the four. Next we notice that A does not affect direction. Obviously not, since if you go ahead it does not alter whatever you do before or after.

The next step is crucial. 'A' operates rather like 1. It does not affect the result. In our next section on algebra we shall see the relevance of 'identity elements'. Let us therefore think of A as if it were 1. Next we look at U. It means we turn exactly round and go in the opposite direction. It makes sense to say $U = -1$. It is necessary fully to absorb these steps before we move on, and make sure they properly model a real situation. We now observe:

$$L^2 = U = -1$$

and this means that L is rather like $\sqrt{-1}$. If we accept $+1$ as a model for 'go ahead' and -1 as a model for 'turn back' then i is a model for 'turn left'.

Is there a catch? Surely we also have

$$R^2 = U = -1$$

which suggests $R = i$. That is so, yet we must agree, with our rule of direction, that $R = -L$. So let us try $R = -i$. All is well. R^2 now does equal $(-i)^2 = (-1)(-1)i^2 = +1(-1) = -1$ and there is no inconsistency. So our model has

$$A = +1$$
$$U = -1$$
$$L = i$$
$$R = -i$$

(It would have been possible to have $R = i$ and $L = -i$; it would also have been consistent, but we here adopt a convention that anti-clockwise turns are positioned so $L = +i$ and $R = -i$)

In this representation, i or $\sqrt{-1}$ is represented by 90° turn to the left. The mysterious square root of minus one is effectively a turn we take many times a day. Finally let us rewrite the table with the new values in. Nothing is altered, we merely replace according to the four equalities above:

	1	i	−1	−i
1	1	i	−1	−i
i	i	−1	−i	1
−1	−1	−i	1	i
−i	−i	1	i	−1

This is the same table, with letters replaced by numbers. The first table made sense in the city grid. The second makes sense if we use our rules for positive and negative numbers and our simple fact $i^2 = -1$. Yet, we say again, the two tables are really the same, or at least isomorphic (the same shape).

We end on the simple statement. Left turns model the square root of minus one.

So number is not simple. This is not surprising, for from it we find vast stretches of mathematics developing. Number is the basis for much of what we do – not number in the sense of calculation, but number as an abstract concept, yet always with practical connections and relevance.

6

Geometry

Intellectual advances take a long time to penetrate widely. If that fictional character, the man in the street, were asked what geometry is about, he might remember the angle sum of a triangle, and that there was a theorem named after Pythagoras. The name Euclid might have stuck. So the general perception is a dim memory of some work done two and a half millennia ago. In this chapter we hope to show that there are geometries other than Euclid's and that they have some external relevance.

The original impulse towards geometry, which literally means 'earth measuring', was purely practical. The builders of ancient Egypt knew that if they used a knotted rope, with twelve equal gaps between the knots, then when the rope was formed into a 3, 4, 5 triangle there was a right angle opposite the largest side. This naturally was an important issue in building.

Euclid not only collected such facts but asked why and when such things happened. The theorem which established the fact that if the square on the long side equalled the sum of the two on the other sides (in this case $25 = 16+9$) then there was a right angle, was credited to Pythagoras. Euclid remains a marvel. Starting with certain things he held to be evidently true, and

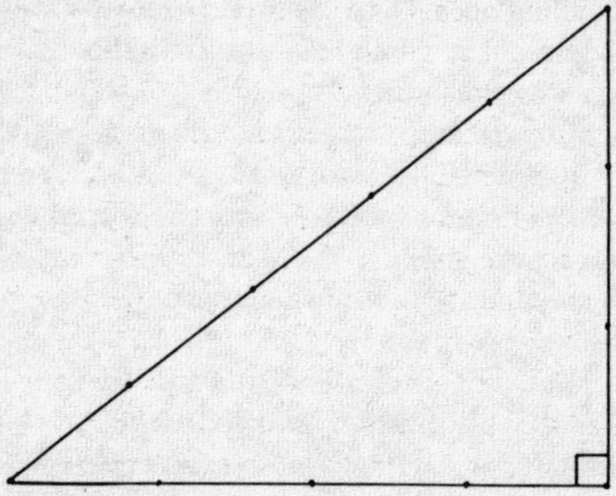

The Egyptian rope-stretchers

- - - - - - - - - ✗ - - - - - - - - - - - - - -

Euclid's fifth postulate

with a number of definitions, he created a structure, step by step in logical reason that stands (although viewed differently) today. It is his method even more than the result that impresses. Religious and political theories have attempted to build structures using the same methodology on different material.

From this mass of work we shall abstract two things to aid us in our further work.

Definition: A straight line is the shortest distance between two points.

Axiom: Through a point not on a line, one and only one line can be drawn parallel to the first.

The second of these two nuggets is known as Euclid's fifth postulate. It worried people; they felt it should be possible to prove it from the other axioms and much effort went into doing so. An Italian priest named Saccheri (1667–1733) believed he had done so. He assumed that the postulate was not true and deduced a series of results which he believed eventually led to a contradiction. Had this been so, it would have proved the postulate, thereby meaning that we did not need it as a separate axiom. No doubt he was rightly very pleased with himself. It would have surprised him to know that his was the first work to show that other geometries were possible, and hence to dethrone Euclid, rather than to free him of all error, as he believed he had done. Let us look at a non-Euclidean geometry, but in a very practical way.

When we fly from London to San Francisco, we may be surprised to find ourselves crossing polar regions. Look on the 'ordinary' map of the world and draw a 'straight' line and it goes nowhere near the frozen North. Do they deliberately go a long way round?

Now take a globe of the world and stretch a string as tightly as we can between London and San Francisco and we begin better to understand the route. It is said that in English the term 'straight line' derived from 'stretched linen', meaning a linen thread. The line between any two places on the globe that marks the shortest distance is part of a 'great circle' – that is one which has its centre at the centre of the globe. Any flat map will distort the picture.

Let us recapitulate. Before we even thought of geometry, man stretched ropes on the ground and made triangles. The Greeks abstracted from these triangles and built an ideal geometrical world, with a thing called a straight line in it that was 'the shortest distance between two points'. They proved that the angles of a triangle, cut off and put together, made up a straight line. The lines came from the real world, were etherealized into having length but no width, and then were supposed to represent some eternal truth in the universe.

Suppose the ancients who made triangles on the earth had made rather bigger triangles. Would they indeed have always had an angle sum of 180°? Back to our modern days. If from San Francisco we travel to Rio de Janeiro

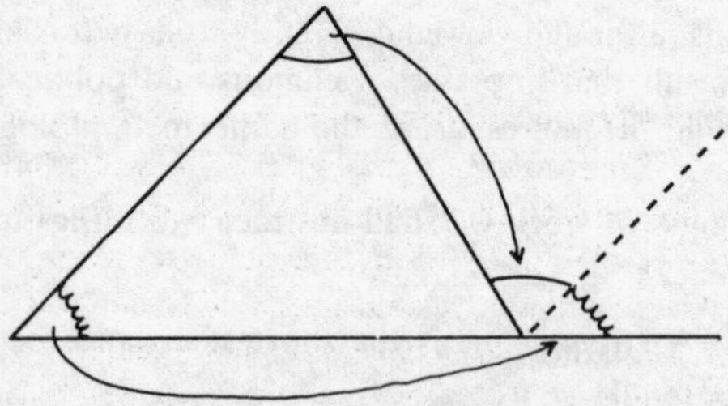

Angle sum of flat triangle

before returning to London, thus completing a triangular tour, do its three angles add up to 180°? They do not. Did we travel in straight lines? We certainly went along the shortest distance between two points. A triangle of this sort on the face of the earth does not have a fixed angle sum. In our diagram on p. 81 we show a triangle running from the North Pole, down the Greenwich Meridian to the Equator, round it for 90° and back up to the Pole. All three angles are 90° – total 270°. Any such triangle has a sum greater than 180°, but its size determines what exact amount it is. The objection will be that these great circles are just that, they are not straight lines, which because of the spherical nature of the earth would go off into space, or go through the earth. This is in fact an assertion that you wish to remain with this mind-picture of a straight line, even though it does not fit the surface of the earth, which is what (mostly) we travel on. So be it . . . for the moment.

It must be accepted that there is a geometry which fits the earth's surface, and that it must be remarkably more valuable for getting about than is Euclid's, unless the journeys are pretty short, when both geometries give results not far apart. As for calling great circles straight lines, that is a matter of semantics, and was certainly in keeping with the shortest distance definition which we lifted from Euclid.

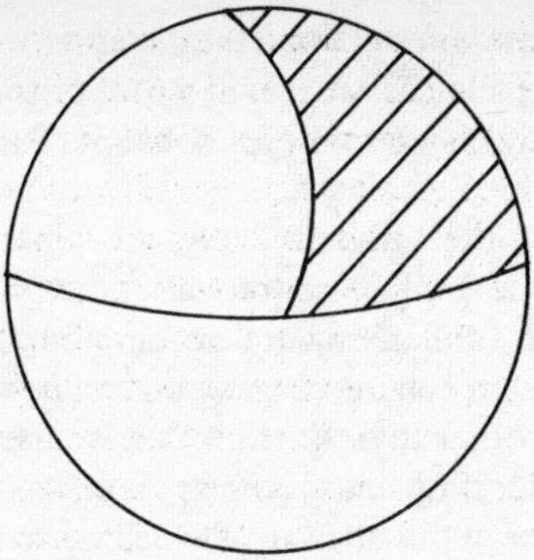

Angle sum of Earth 'triangle'

On the surface of the earth there are many other lines, or curves, such as the lines of latitude. These are not 'straight lines', because they are not shortest routes. The only line of latitude which is a great circle is the Equator. If two cities happen to be on the same latitude, we do not fly along it to get from one to the other. We have, on the earth's surface, used Euclid's definition involving shortest distance. Look now at his fifth postulate. Start with a 'straight line', for example the Equator. Take any point not on the Equator and put a 'straight line' through it. Since the great circle has its centre at the centre of the earth, you will soon find that no matter where you draw a line through the point it will cut the Equator (twice).

In the geometry of a sphere's surface, you can draw no line parallel to a given line through a given point. This was, effectively, one of the propositions that Saccheri had started with when he sought to prove the postulate by contradiction. What he produced, among other things, were theorems that were true on the earth's surface (surely a matter of some relevance) but not in Euclidean geometry.

The geometry we use in getting about the world is non-Euclidean. There is an amusing irony in this, since geometry is supposedly 'earth-measuring'.

The uncertain feeling lingers, because the Greeks have left us this mind-picture of a straight line, heading out for ever, so let us follow it. Even though we finally got clear (nearly all of us) that the earth is not flat, it would be satisfying if these lines fitted the larger picture and if three-dimensional space could be made up of sets of parallel lines, going off in three separate directions, all at right angles. We feel they must be right somewhere in the real world.

Newtonian mechanics and the universal system of gravitation assumed a space of a Euclidean sort. This was convenient, for it meant we could rely on a lot of geometry we already knew, with the planets travelling in elliptical orbits and things generally behaving like conic sections. It worked, and we fell into the same trap yet again. It worked when the distances were not very big. When we last spoke of short distances, we meant a mile or two. We now mean sizes rather bigger than the solar system. Beyond that our Euclidean geometry does not seem quite right.

Einstein's view of the universe, developed in the general theory of relativity, needed the whole of space (or spacetime) to be a bit bent. We are obliged to make the very difficult mental accommodation, of space behaving rather like the surface of a sphere – but we do not now have an inside. Once again, it is a geometry other than Euclid's that we use.

We began to wonder if Euclid's geometry is any use. The fact that it is not an accurate model either of the surface on which we are largely constrained to move, nor of the universe in which we live, does not make it invalid. It has become, as much mathematics is, a creation of our imaginations, but it remains extremely useful in practice. The whole range of architectural and engineering drawing uses Euclidean geometry, and perfectly sensibly so. If you build a very long bridge, you may need to account for the Earth's curvature, but why go into spherical geometry, with its decidedly awkward triangles, when Euclid is near enough?

The point is established that Euclid is not the only geometry, but it leaves us unsure exactly what geometry is. The Greeks felt it described the world in a very deep manner. Their mystical sense also led them to believe that, for example, the circle was the 'perfect' shape and had some eternal meaning in consequence. Certainly their geometry is a powerful creation, and as we have said, we have no need to justify it in worldly terms; it does not have to fit, it has to be.

We are free to create geometries with what axioms we like, and it is a surprise to most to know that there are many geometries, each with its own internal consistency. Some we can find models of, as in that which fits the surface of sphere, but it is not necessary that we should be able to model it, nor need it be of use in any real situation.

Our next geometry opens with a problem. Along a straight road lamp posts occur at regular intervals. We could represent this as a line with dots spaced equally along it, marking the feet of the lamp posts. Even doing that

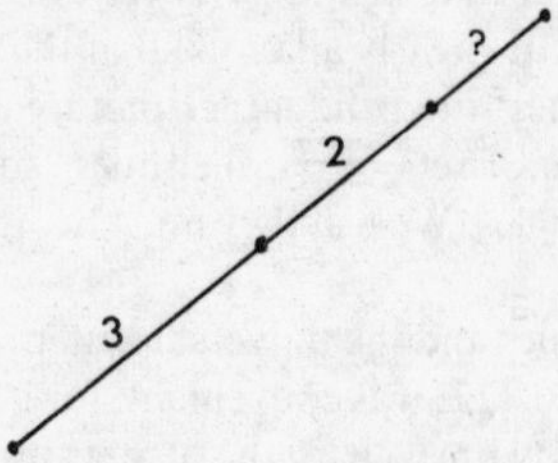

Lamp posts on a street

takes us well into abstraction. For practical purposes we assume the surroundings Euclidean, though we do not wish to present it as it is, but as it looks. We are attempting to draw in perspective. The diagram shows the distances between the feet of three successive lamp posts, which happen to measure 3 units and 2 units. We know they are really equal, but we allow for foreshortening.

An artist would unerringly put the foot of the next post in the right place, and would then proceed to fill in more of them. A trained eye can be a very effective tool. Not everyone would be so successful. A few (very few) make the next gap bigger than 2. Rather more, but not many, make it equal to 2. By far the majority think that it only looks right if smaller than 2, and in this they are right. Yet there is some variation in how much shorter they make it.

The mathematician's task is to calculate where the next point is, and all the points after this. The first inclination is to reduce in the same proportion, and take $\frac{2}{3}$ of 2. But if we draw it, it does not look right. The numerical relation lies deeper and we describe it in the Appendix (pp. 232–3).

In Renaissance times the interplay between art and mathematics was intense. Straight lines running to 'vanishing points' are manifest in Leonardo, Raphael and others, but most particularly in Piero della Francesca, who fused art and mathematics to a very marked degree. Our starting problem is a very easy example of what the artist in pursuit of realism had to tackle. The geometry involved in perspective drawing became known as 'projective' geometry.

A projection is a number of lines coming out from a point. Dürer and others looked from a single point, through a flat screen at the object they wished to paint. If every point was marked on that screen, lines ran from all parts of the object to be painted to the eye, leaving a flat picture on the screen. A slightly simplified version is shown in our diagram. Here the real object (a

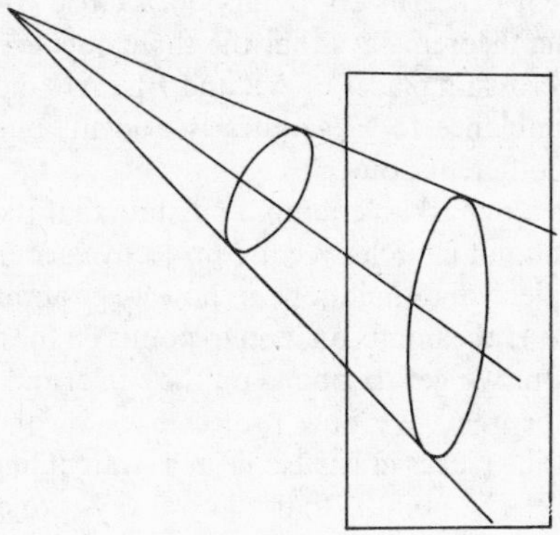

Projecting a circle

circle) has its shadows thrown on a screen, this time behind the object. The result, an ellipse, is the projection of the circle.

Projective geometry is not just about lamp posts or circles with elliptical shadows. It looks at just those things which are not affected by that change (or transformation, as we say). The distance between lamp posts did not stay the same, nor were the angles between them and the road drawn as right angles; they were so in reality, but it would have looked wrong. This means that lengths and angles, which seem so essential to Euclidean geometry, do not stay fixed in projective geometry. Yet something must stay fixed, for a photograph is immediately recognizable by those who know the person or the place it represents.

We will stay in geometry and look at two projective theorems, named respectively after Desargues and Pascal. Oddly, Desargues is easier to prove in its three dimensional version rather than its plane one. The two shaded triangles are 'in perspective'. This means that viewed from P one triangle exactly covers the other, though they are not the same size or shape. Another way of looking at it is to see the lower one as the shadow of the upper with a light at P. (An interesting side issue is whether, given two triangles, you can always put them in perspective, no matter what size and shape they are; this is left to the reader.)

Desargues' theorem states that if you extend corresponding sides of the triangles until they meet (PR and XY extend to meet at A; QP and ZX to meet at C; QR and ZY to meet at B) then the three points A, B and C all lie in a line. It takes a very practised eye, that of the mathematician or the artist, to get this diagram clear in one's mind. The power of visualization is spread unevenly among the population. The proof of the theorem is given on p. 233.

The point of the diagram is that it highlights a number of points about drawing in perspective, and that within results such as this lies the underlying theory of how to draw.

Pascal's theorem concerns six points on a circle. In our diagram the points are A–F. The theorem says that the three dots are in a straight line. These dots are the crossing places of AE and BD, AF and CD, and BF and CE. A family resemblance to Desargues is evident, but we shall use the theorem to make a different point.

We now make a remarkable jump, and state that it would still be true were it an ellipse and not a circle. We use projective geometry to prove this. Our second (incomplete) diagram shows us how we go about it. We project the ellipse into a circle, and the points on it on to points on the circle. So A goes to A′, B to B′ and so on. We get six points on the circle, and can then draw the result of Pascal's theorem. We now project back on the ellipse when the corresponding crossing places must also lie in a straight line, and so the result is true for the ellipse. The point is that a line always projects on to a line. If three points are in a line in one diagram they must be in the other.

Properties such as the fact that three points are in a line are called

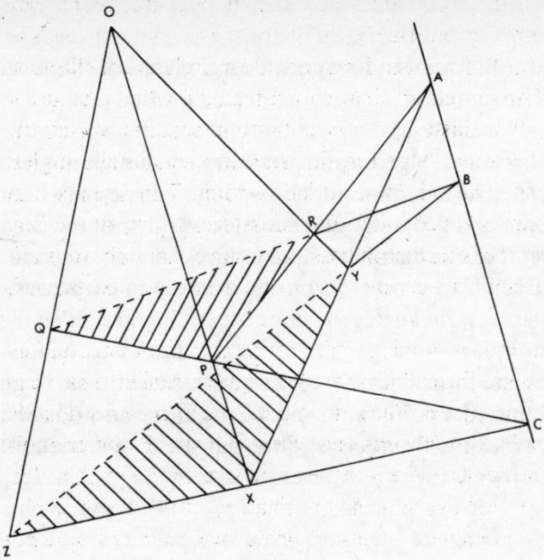

Desargues' theorem

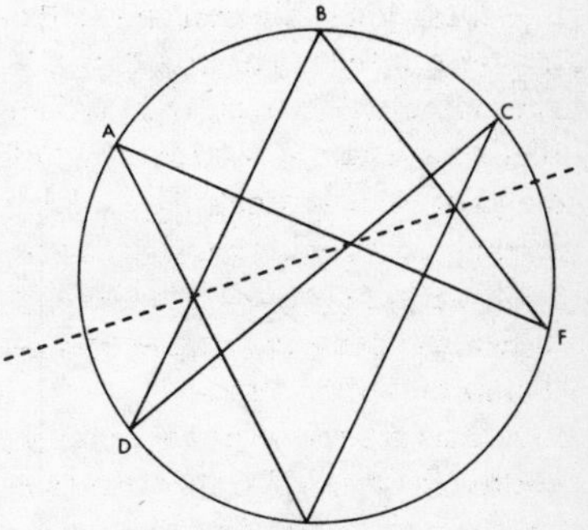

Pascal's theorem (1)

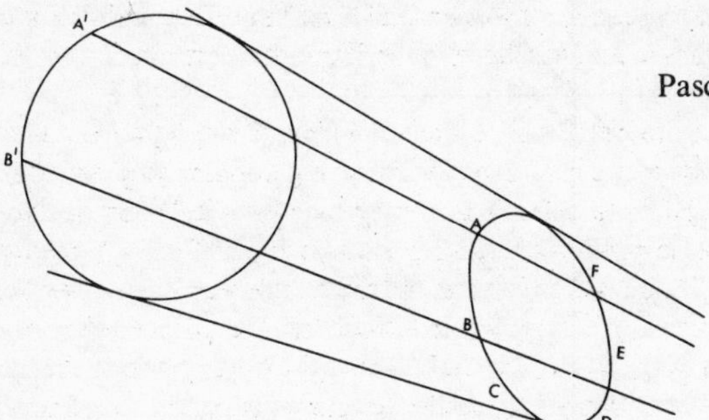

Pascal's theorem (2)

'projective' properties, because they are not affected by projection. In a photograph, something that is in reality a line will appear as a line. Projection has been a powerful mechanism for extending geometrical results of this sort, from circles to any other conic section.

It was Descartes who first fused algebra and geometry in what we now see as a very normal connection. The invention of 'co-ordinate geometry' gave a great impulse to both algebra and geometry and revealed another of the deep connections in mathematics that can lie hidden for centuries and look obvious when made explicit. It is Euclidean geometry, but interpreted in equations.

Descartes drew two lines at right angles to one another; they are called the x and y axes. It is commonplace nowadays in the mass media to express numerical material in graphical form against two such axes, but we need to go slightly further. We have labelled a point (x, y) and we should regard this as a movable point. It does not move without constraint, but within a constraint contained in an equation. If we write

$$y = x$$

we ask that the distances the point stays from each of the two axes are equal. Our geometrical intuition tells us that the point can now run up and down the

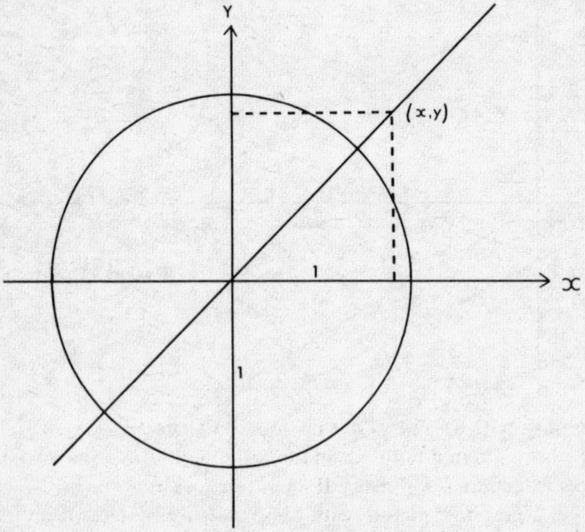

Cartesian coordinates

line at 45° to both axes, as shown in the diagram. We thus have an equation and a line which are effectively the same thing. This is really an extraordinary state of affairs. Both equations and geometry had been studied in great depth and with intensity without this connection being apparent before Descartes.

Drawing a circle using an equation is more difficult, but

$$x^2+y^2 = 1$$

is a connection between x and y that is only true on the circle round the origin of radius 1.

Not only do we draw pictures using equations but connections in the pictures match connections in the equations. Two simultaneous equations in algebra, such as these

$$x+2y = 7$$
$$2x-5y = 5$$

have a simple answer for each of x and y: $x=5$, $y=1$. In the picture, the equations become lines and their solution the point where the two lines cross.

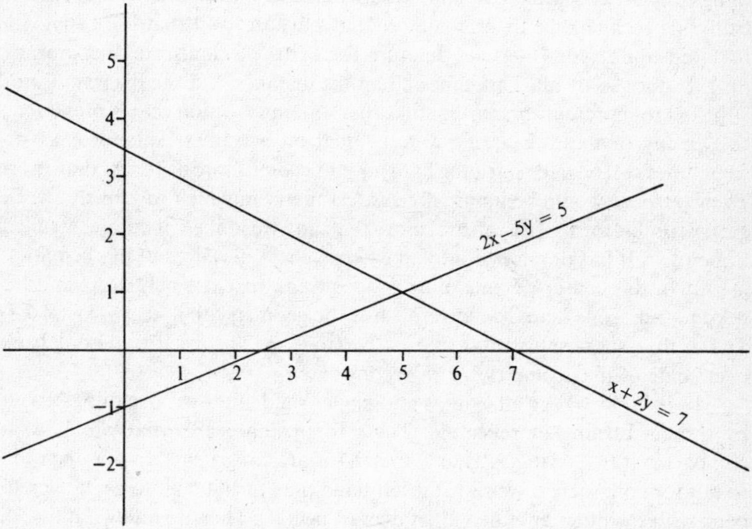

Equations with xs and ys but no squared terms (or xys) are all straight lines. If we have squared terms, as in the equation of the circle we gave above, we get a conic section of some sort. If we put y equal to some expression with x^3 in it, we get a graph or picture with two bends in it – and so on. For every situation in the algebra a corresponding situation occurs in the geometry, and vice versa.

Certain geometrical theorems are difficult if we only use geometry, and fairly routine if we move into algebra. Against this, geometrical insight can help in our understanding of algebraic forms. It is a happy liaison.

Consider public transport systems around the world. Whether we be in Moscow, Paris, London or San Francisco, we will be offered diagrams; whether it be the Metro or the BART, the diagram consists of lines with dots on them, crossing each other. That is the simplifed, mathematical abstraction of what they are. The lines are different routes, the dots stations, and the dots on intersections of the lines the places where we change.

The lines are sometimes straight and sometimes not. The person who devises the plan (it is not technically a map) bends lines if it is convenient to do so, whether that happens on the real system or not. In stretches of the lines where there is no special reason for arranging the dots, they are probably equally spaced, though again that it not true in the real world.

The diagram, therefore, in terms of any geometry we have looked at so far, is a mess. It does not remotely represent the whole picture as it is. Yet it perfectly fulfils its purpose. The traveller normally requires to know only two things, the order in which the stations are reached, and where you can change from line to line. The diagrams meet that purpose, and that is all they are required to do. The Inter-city diagram for British Rail also tells us journey time. It does so by marking times along the distance, but the diagram is not adjusted to represent these. It is still order and intersection that is preserved.

If any of these diagrams were printed on a rubber balloon or as we sometimes see, on a tee shirt, the lines might become more distorted than ever. Their purpose would be unimpaired. A geometry in which we can still work even with this form of distortion, more than any we looked at before, is called 'topology'. It has developed into an extensive and new area of mathematics, mainly in this century. Some of its propositions are extremely deep, and far beyond our scope here. In our terms it is the geometry that keeps least fixed. That is the issue we must address ourselves to – 'What stays fixed?' Look back over some of the geometry we have considered.

In Euclid we keep all lengths and angles rigid and we move things about by sliding, turning or reflecting. These for instance are what we allow in comparing if two triangles exactly match (or are congruent). In our diagram, to test for congruence, we slide the left hand one, turn it a bit to get it back to back with the other and then flip it over (reflect it). Then we know if it fits. In Euclidean geometry we know results such as that the fixed angle sum of a triangle is 180° or, later on, that the angle in a segment of a circle stays fixed as we move the point round, keeping a chord fixed. These are some of the things that stay fixed when we allow the three sorts of motion we mentioned. The results of Euclidean geometry are excellent for all building and engineering, at least where we are dealing with solid rigid chunks of material like bricks or concrete slabs.

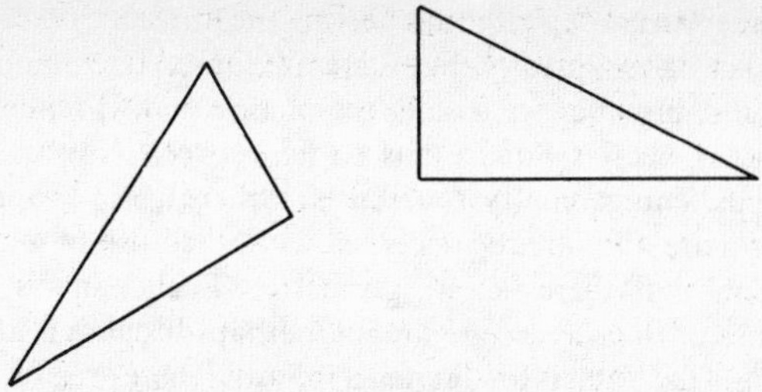

Euclidian congruence

In our geometry of the globe we attach a different meaning to straight line, and the fixed angle sum of a triangle is with us no longer. It is an interesting exercise for the reader to work out what features of two spherical triangles need to match for the whole triangle to match; the result is not the same as in Euclid. Yet length and angle still matter – ask any navigator. The geometry is different, yet important where it is appropriate, as in world travel.

As we go out into space and view the cosmos, again we have a geometry, relevant to the fusion of space and time but, like the geometry of our own globe, non-Euclidian.

Projective geometry finally breaks with the fixity of length and angle. The solution of the lamp post problem given in the Appendix shows that some underlying numerical concept is kept, but we can no longer rely on measurement. If, of course, in a diagram two curves cut once, then they do in any projection of that diagram (if we bring infinity in – a remark for the mathematician). But more is preserved than that. Straight lines remain straight lines, but also a conic remains a conic.

Here we can bring in a comment from coordinate geometry, which is in fact Euclidean, and not different in the sense of the geometries we are discussing at the moment. All conics in coordinate geometry can have equations with x^2, xy, y^2 in them, but no cubic term. The conics therefore appear as a neatly defined group in coordinate geometry. So, too, do they in projective geometry. A conic projects to be a conic. In fact the original cone from which they are seen as sections (diagram on p. 161) can be thought of as the projection from its top point. Projective geometry is relevant and appropriate in representing a two- or three-dimensional reality from a point. In a sense, projective geometry is our basic perception of the world. The picture an eye has of the lamp posts does not have them at right angles to the pavement. Our mind tells us it is so, but not till we walk near to the lamp post does the picture on our retina give us that right angle.

Finally in topology or 'rubber sheet' geometry we destroy even those categories. We can no longer speak of straight lines, for in this geometry we could bend one if we chose, without affecting the information carried. In this geometry we can distort greatly, but there are limits. If a point lies between two other points, however we distort the picture, it will remain so. This is fortunate, since Leicester Square has to lie between Covent Garden and Piccadilly Circus. This geometry, too, is useful where it is appropriate. If order and intersection are all you are concerned with, then use topology.

The notion that there is one geometry – Euclidean – is deep-seated, perhaps because of the way we are taught, perhaps because it looks right for the world around us. Yet it is too limited a view and there are everyday reasons why we should not hold to it.

The discussion of various geometries now allows us to offer an idea as to what geometry is. We have certainly moved beyond 'earth measuring'. It seems in each geometry we are allowed to do certain things, from the restrictions of Euclid to sliding, turning and reflecting, to pulling about the rubber sheet (provided we do not tear it). According to these changes that are permitted (transformations) certain things stay fixed: in Euclid, the angle sum of a triangle remains constant; in projective geometry a conic is still a conic; in topology the points stay in the order.

We are concerned with what stays fixed when we permit certain transformations. Geometry is the study of what does stay fixed under certain changes. Or to give it the mathematician's gloss (which will not help a bit):

> 'Geometry is the study of invariance under a certain set of transformations.'

7

Algebra

There is a mystique attached to mathematics that cannot but be harmful. Some of it has been encouraged by mathematicians. The school of Pythagoras indulged themselves in much mysticism and formed a sort of priesthood. It is always tempting to belong to a closed sect; priests always have power. The feeling about mathematics remains, and it is algebra that is often seen as the introduction to the 'higher mysteries'. Yet there is no mystery about algebra.

It is a rather odd name. The original Arabic meant the reunion of broken parts, the restoring to normal, and the equating of like with like. Certainly in what we now mean by algebra we are concerned with equating like with like; equations are a significant feature of algebra and one with which we might well start, but there is perhaps an even easier entry point.

When we first get used to adding whole numbers together it does not take us long to gather that whether we add 2 to 5 or 5 to 2 does not matter as far as the result is concerned. If we wish to explain this to someone who does not know we might say

$$2+5 = 5+2$$
$$\text{and } 7+3 = 3+7$$
$$\text{and } 4+3 = 3+4 \text{ and so on} \ldots$$

adding perhaps something like 'It doesn't matter which number you take first when you are adding'. The rather straggly explanation giving examples is unsatisfactory, since examples cannot cover all cases. The language statement is better, but neatness and precision is much better conveyed by

$$a+b = b+a$$

where we all have an agreement that a and b stand for numbers. It is worth thinking about what has been done here, for if we get it clear, later issues will not get clouded. There is no special reason why we should choose letters. We could not use numbers and still convey that the rule applied to all numbers, but we could have used a symbol which we invented for the purpose to express a general, non-specific number. However, having the alphabet symbols to hand, mathematicians used them. The trouble is that there is no connection between the mathematicians' meaning and the written language meaning, and that is confusing. This is a difficulty introduced by a decision of an arbitrary nature; it is not a conceptual problem.

There is a conceptual problem, and it needs to be sorted out here or it will create trouble. We all have the feeling that a and b are merely disguises for numbers that someone actually knows. This will prove valid when we look at equations, but is not valid here. The idea that 'a' can be any number at all (and so can b), and therefore is both 'any' number and 'all' numbers, is one we must accommodate in our thinking. The result we have shown applies always, and we are free to put such values as we choose in place of a and b.

We have another problem. The statement looks like an equation, yet it is not. It is what we call an identity, simply because it is always true. An equation, as we shall shortly see, is true only for some values, or numbers, that we put in it. For the sake of precision, we sometimes write

$$a+b \equiv b+a$$

where the three lines represent something stronger then equality.

Let us given an analogy: there is always a risk in this, since analogies seldom fit exactly. Put on a balance two identical packets of sugar. They will not only balance, they will look exactly the same. Now remove one and pour nails into its scale pan. We can still achieve a balance – the sugar and the nails have the same weight when we have done so – but they are no longer identical.

The identity tells us one of the rules about working with numbers. There is no way we can 'work out' a and b. There are other rules, all of which seem commonplace enough

$$a+(b+c) \equiv (a+b)+c$$

This says, in effect, that if you are adding up three numbers it does not matter which pair you add together first. The brackets on each side of this identity group together those which have been added first; it is a convention, not a reason, that says we do the brackets first.

These two rules are called the 'commutative' and 'associative' laws. The words are well chosen, and reflect the laws they name. There are two very similar rules for multiplication

$$ab = ba \quad \text{(commutative)}$$
$$a(bc) = (ab)c \quad \text{(associative)}$$

The first says that you can multiply either way round (you can commute) and the second that it does not matter which pair you 'associate' in doing the first multiplication.

We have introduced another convention. When we write ab we mean $a \times b$. There is no reason why we should drop the multiplication sign; it has merely become the practice. Practices of this sort make life difficult for everyone. Let no one try to tell you that mathematics, and particularly its symbolism, is clear, concise, and unambiguous. It is a mess. We have to come to live with it as we have had to accept English spelling. To illustrate the mess look at these three conventional notations:

$$2\tfrac{1}{2} \text{ is } 2+\tfrac{1}{2}$$
$$25 \text{ is } (2\times 10)+5$$
$$2a \text{ is } 2\times a$$

In every case we have written a symbol next to the numeral 2. In every case it means something quite different. We will never tidy this up, it is too late, but it helps to recognize that it is the mathematicians who are in error, not those who are attempting to understand it.

There is one more rule that we need to look at

$$a\,(b+c) \equiv ab+ac$$

This says that if you add two numbers and then multiply by another the result is the same as multiplying each of the two numbers separately by the third and then adding the results. Notice that the wording we need to use is already becoming far more clumsy than the symbols.

This law is called the 'distributive' law – the 'a' is as it were distributed across the b and c. It helps to put some numbers in to see what is meant. Say we put $a=3$, $b=4$, $c=5$, then if we work out $3(4+5)$ that is 3×9 or 27. On the other hand $(3\times 4)+(3\times 5)$ is $12+15$ and that is also 27. Putting any number of numbers in is not going to prove anything – the law expresses how addition and multiplication relate one to another, and is a starting point rather than something we establish.

These are the basic five rules that govern the basic operations between numbers. We have not looked at subtraction and division, for they are reverse processes (in mathematics we usually say 'inverse' processes). A mass of school algebra depends on altering some algebraic 'expressions' into simpler forms using such rules. An expression is a series of algebraic terms with mathematical signs. A simple one is

$$2a+5b$$

It is totally natural to try to make this look simpler, even if you did get slapped down at school for it. After all, if we knew a and b we could work it out and get a single number. But while a and b are still totally general and non-specific we can do nothing with this expression, for there is no connection between a and b. However

$$2a+5a$$

does simplify to 7a. (Do not get conned by the trick of saying that a and b are apples and bananas and you cannot add. They are numbers.) This expression simplifies because

$$2a+5a \equiv a(2+5)$$

This is the distributive law at work. Our right hand side now becomes $a \times 7$ or 7a (commutative). It is making a mightly meal of it, but understanding is essential. Working by remembered rules will leave one totally bereft later on.

Let us now do some joining up of broken parts, as algebra purports to be.

$$\text{Simplify } 3a+2b+4a+6c-2a+c-b$$

It looks a bit fragmented, but putting like with like we have

$$a(3+4-2)+b(2-1)+c(6+1)$$
$$\text{or} \quad 5a+b+7c$$

Neater, and more unified; not especially clever or difficult to do.

Algebra of this sort is known as 'generalized arithmetic'. It works with non-specific numbers, but they all obey the ordinary rules for number. Since we do not specify them, we generally do not work out as far as we would in arithmetic.

A great deal of time is spent in our mathematical education in knocking one algebraic expression into another identical, but perhaps simpler looking one. It is slog work, requiring practice but nothing creative or really thoughtful. Algebraic manipulation plays the same role in algebra that calculation does in arithmetic, necessary but rather boring and better done by machine, to leave humans free for thinking.

The manipulation depends on the rules we have quoted. A proper understanding of these is all the understanding needed. That does not mean there is not a lot to do if you want to become adept. It is quite a lot of work in getting from being able to do 13×53 to doing $856{\cdot}786 \times 631{\cdot}832$ – but it needs no more understanding. The same is true in algebra.

Solving equations is quite another thing; and rather more interesting. If we write

$$x+2 = 5$$

we are doing something quite different from saying

$$a+b = b+a$$

In the second, the two sides are always equal. In the first, if we regard x as a general, non-specific number, the statement is mostly not true. In fact it is only true when $x=3$, as everyone has been leaping to say. Return to the standpoint of regarding the equation as a provisional statement

$$\text{`}x+2 = 5\text{'?}$$

It has become a sentence, in inverted commas, whose truth or falsity is to be discussed. Compare it with:

'The King of England had his head chopped off'

Is that statement true? 'The King of England', though it sounds like one person, is non-specific. There is not, as with x, an infinite range of possibilities but there are quite a number.

This language statement is true if

'King of England' = 'Charles I'

but is untrue for other values of 'King of England'. Our equation is true if $x = 3$ and not true for other values of x. The two statements have an underlying similarity.

Again, we seem to be making a meal of it. Yet it is necessary to establish when we are working with expressions and when with equations. Look at a slightly different equation:

$$2x + 7 = 3x - 5$$

We have two different expressions on each side and they form an equation. Let us not rush into half-remembered rules about 'Change the side and change the sign'. This sort of rule only increases the mystification. Rules are only worth using if you know the reason for them (and that is not only in mathematics). So let us play about putting some values for x in. Doing it in an orderly way helps:

x	$2x+7$	$3x-5$
1	9	-2
2	11	1
3	13	4
4	15	7

Are we now getting a 'feel' for it? One column starts with 9 and goes up in 2s, the other starts below zero but goes up in 3s. Shall we jump a few steps?

10	27	25	
11	29	28	
12	31	31	Bingo!

This is better.

So we found a number of cases where the statement was not true and just one, $x = 12$, where it was. Look at the numbers and see if you think any other number will do. Obviously not?

Our last chapter on geometry showed that we could link algebra and geometry to the advantage of both in coordinate geometry. Our picture on p. 96 does just that. We have used the numbers in our columns to show how big each expression is for different values of x in a pictorial way. The visual impact of where they cross is very powerful.

The rules we mentioned come from a different idea. We know that if a pair of scales balance they will still balance if we add or subtract the same from each side.

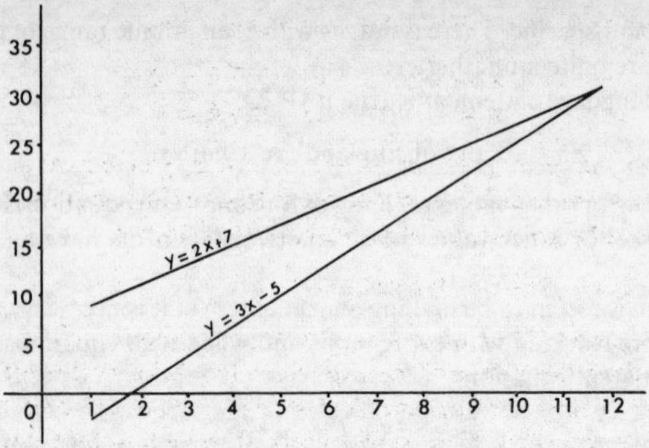

Algebra and geometry: simultaneous equations

If $2x+7 = 3x-5$

we can say it is the same as saying

$$2x+12 = 3x$$

because we have added 5 to each side. Now we can see the answer is 12, but if we want to be pedantic, take 2x from each side

$$12 = x$$

The crazy rule 'change the side, change the sign' amounts to the same thing, but we are not machines; we must know that what we are doing is reasonable.

Understanding comes from seeing it in different ways. Adherence to one method is simply a crutch to reaching the answer.

Provided that there are only xs and not x^2 terms, there will be a single answer. But supposing we had

$$x^2-5x+6 = 0$$

(x^2, remember, is x times x). Again, regarded as a statement about a non-specific number x, it is not often true. Take the expression x^2-5x+6 and see how it works out for different values of x, rather as we did before.

If $x = 0$ then $x^2-5x+6 = 6$
$x = 1$ $\quad 1-5+6 = 2$
$x = 2$ $\quad 4-10+6 = 0$

We seem to have struck lucky, and hit the answer rather quickly. The inclination is to stop – but let us go on, because we have not got a pattern.

If $x = 3$, $x^2-5x+6 = 9-15+6 = 0$

and we seem to have another value that makes it true

$$x = 4 \qquad 16-20+6 = 2$$
$$x = 5 \qquad 25-25+6 = 6$$

and a pattern has emerged . . . 6 2 0 0 2 6 . . . it is symmetrical at least.

Experience in calculation in arithmetic leads us into the false belief that a question has a single answer. Here we have the question

'When is $x^2-5x+6 = 0$ true?'

Answer: 'When $x = 2$ or when $x = 3$' as we have shown.

Let us draw the picture of this expression for various values of x. We plot the x value against that symmetrical string of numbers, and the picture reinforces the symmetry. If we were concerned about how far below the x axis the curve goes we can work out the expression for $x = 2\frac{1}{2}$

$$x^2-5x+6 = 6\tfrac{1}{4}-12\tfrac{1}{2}+6 = -\tfrac{1}{4}$$

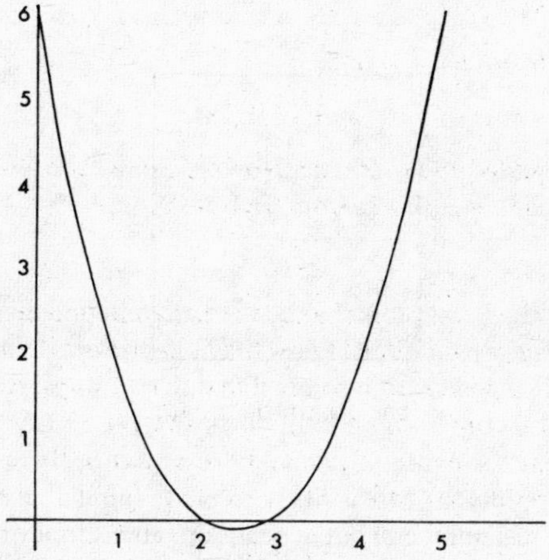

Equations: values of x

The whole curve (parabola) shows us that the expression can take many values. Just twice is it zero, and the curve is convincing as to why it is twice.

We can now give a real-life situation where two solutions are natural – and attach it to this graph. Turn it upside down and we have the path of a stone in flight. Suppose it moved in this symmetrical curve and reached a height of 12 feet at its apex. Then if we ask 'When is it 8 ft up?' there are two answers, whether we are interested in the time or the horizontal distance out from the start. This is a natural problem which has two answers.

We have tackled it by trying numbers and by drawing a picture. Can we try the balancing trick again? We can try, but it does not work. Because there are both x and x^2 terms, it will not be easy to get on its own, as we do with a simpler equation. For a few equations, of which this is one, there is a method that will work. If we rearrange the expression $x^2 - 5x + 6$ according to our rules we can get it to this form $(x-2)(x-3)$. This means that we have two numbers multiplied together; the only way that they can come to zero is for one of them to be zero. The first is zero if $x = 2$ and the second zero if $x = 3$, and that means they are the answers.

We have seen three ways of finding the answer to this 'quadratic' equation:

(1) Filling in numbers and spotting a pattern;
(2) Drawing a picture (a graph) of the expression;
(3) Writing the expression as two numbers multiplied together.

The word 'quadratic' comes from quadrate, or to make square. This is used because there is a term in x^2 and the geometric connection is obvious.

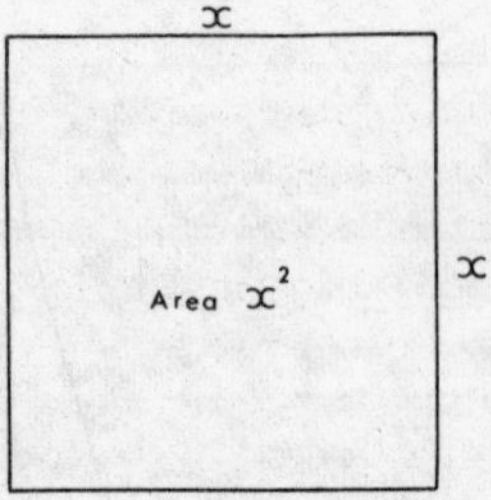

The quadratic square

Of these methods, the first is likely to prove unreliable, the second will always work (for real numbers) but depends on our accuracy in drawing. The third depends on being lucky with the equation. Mathematicians are always looking for methods that will work generally, not just for specific numbers, and there is a way.

We shall use the equation we have been working with to illustrate this. Our picture showed it is balanced about the point where $x = 2\frac{1}{2}$, and there is a finesse based on this symmetry. We say that $x^2 - 5x + 6$ could be written $(x-2)(x-3)$. Some generalized pictures for this can be found in the Appendix p. 234. If instead we looked at $(x-2\frac{1}{2})(x-2\frac{1}{2})$, which is not the same expression, but somewhere near and connected with the $x = 2\frac{1}{2}$ balance point, we find

$$(x-2\tfrac{1}{2})(x-2\tfrac{1}{2}) \equiv x^2 - 5x + 6\tfrac{1}{4}$$

This is not an equation, but a rearrangement of terms. If we write

$$(x-2\tfrac{1}{2})(x-2\tfrac{1}{2}) = \tfrac{1}{4}$$

that is an equation and the same one as before, with $\frac{1}{4}$ extra on each side. The point is now that we can square root each side and get

$$x-2\tfrac{1}{2} = \sqrt{\tfrac{1}{4}}$$

Now the number which multiplied by itself gives us $\frac{1}{4}$ is $\frac{1}{2}$. We may have expected the square root to be smaller than the number, but that is only true for numbers bigger than one. The square root of 4 is 2, but of $\frac{1}{4}$ it is $\frac{1}{2}$.
So we can write

$$x-2\tfrac{1}{2} = \tfrac{1}{2}$$

giving us the answer 3.

But, in our work with negative numbers, we saw that $(-\frac{1}{2})\times(-\frac{1}{2}) = +\frac{1}{4}$, so there is another square root, $(-\frac{1}{2})$
So we can write

$$x-2\tfrac{1}{2} = -\tfrac{1}{2}$$

giving us the answer 2.

We have now added a fourth method to our armoury, but although it may sometimes prove slightly more complicated in its numbers, this way always works. It is called 'completing the square'.

What yields understanding in mathematics is not the detail, however, but the policy. The policy here is that somehow we must square root to get the x^2 down to x. We show in the Appendix (p. 235) how this policy, applied in general, can knock a quadratic down to a 'linear' equation. It leads to that well-known formula

$$\frac{-b \pm \sqrt{b^2-4ac}}{2a}$$

which generations of schoolchildren have committed to memory.

The natural instinct of a mathematician when he or she has solved one tricky problem is to look for something harder, and in mathematics that is never far away. They set about equations such as

$$x^3+2x^2-x-2 = 0$$

This is a cubic equation, because in the first term we have x times x times x and that is a cube. We can go about it the same way as before. Put in a series of values of x to see the pattern. Make this much clearer by plotting points on a graph (see p. 100) and even by trying to rearrange the expression x^3+2x^2-x-2 in a way that has several numbers multiplied together. If you work at this you will find that all methods can succeed. But what mathematicians wanted was to solve

$$ax^3+bx^2+cx+d = 0$$

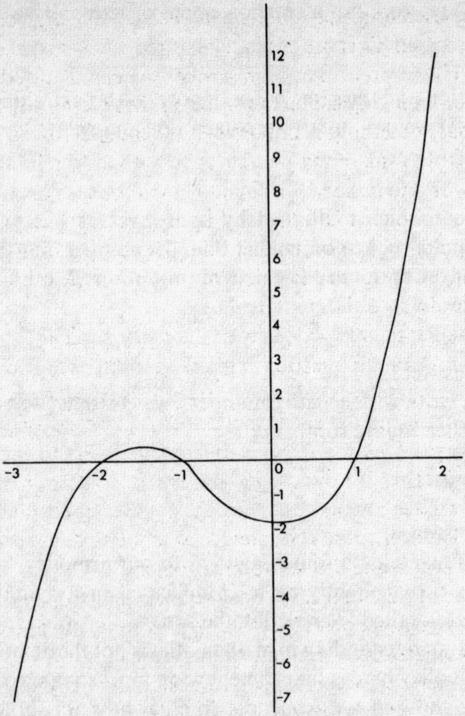

$y = x^3 + 2x^2 - x - 2$ in graph form

where a, b, c and d were general, non-specific numbers. Again, we needed a formula, like the one for the quadratic. A method was found, similar in idea, that reduced the equation to a manageable form. The formula looks worse, but complication is not important, once we have the underlying idea.

As soon as mathematicians had got their teeth into the problem and found ways of solving cubic equations it was naturally thought that they would, with similar methods, be able to solve equations in x^4, x^5, or whatever power we liked. Complications were expected, but not fundamental and conceptual difficulties. The quartic or fourth degree equation succumbed, but the quintic or fifth degree simply did not. Attempts to reduce it to a lower degree resulted in its coming out at a higher degree; the quintic was a slippery beast.

It became one of these problems where well-known mathematicians received 'solutions' from cranks all over the world. One such 'solution' was discovered by a brilliant young mathematician named Abel (1802–1829), a

Norwegian who died tragically young and lived a life of crushing poverty. No sooner had he sent off his solution than he found a flaw and realized it was no solution at all. His response was remarkable: he demonstrated the impossibility of a solution in a single expression using these letters a, b, c, d . . . . To demonstrate that something is impossible is a tour de force. This one proof sets him in the front rank of mathematicians, though he left much else for the mathematical world to ponder. Originality lies in conception, not in the power to manipulate symbols. It is a demanding task to try to solve equations of increasing difficulty – but an obvious one. There may be much subtlety in finding ways of doing so, yet to think of proving impossibility and then to conceive how to do so is different again.

Abel's work led to great extensions in algebra, too deep to consider here, but there is an interesting question remaining in the whole enterprise. Why exactly did we set ourselves these constraints for the solution – that it had to be a finite formula using the coefficient a, b, c, . . . ? Newton and others had found simple ways of getting closer and closer answers to the 'roots' of any equation, no matter what powers of x there were. With a micro beside us, a short program will get an answer to a great number of decimal places to any such equation. The constraints certainly lead to interesting new mathematics; yet the engineer wants the answer to his problem.

Let us pause at this stage and consider the content of traditional algebra that most of us learn at school. It looks at expressions with letters in them, and how they can be rearranged. Endless hours are spent (not very profitably) in this exercise. Some expressions we call formulae – usually when they express one particular result. If you drop a stone, it falls $16t^2$ feet in t seconds. $16t^2$ is a formula (and an expression). If you want to know how far it falls in 3 seconds, you put in 3. Then 3^2 is 9 and 16×9 is 144, so we know it falls 144 ft. – and that can be practically useful. Another formula is the solution of the quadratic equation we have just seen. Given any quadratic equation, that sorts it out.

The other major task has always been to solve equations – but not beyond quadratic; we would find few Abels among the population. Put like that it is very little – and this is true if we fully grasp what it is all about.

The real-world task is to abstract from a situation the equation which relates to it. The height of a stone as we hurl it away might conform to this:

$$y = 16 - 5x - x^2$$

but the way we get from the real situation to this mathematical expression of it is very important. We shall discuss a simple case of a stone in flight in our chapter on calculus, which will give some flavour of how we move from a real world to a mathematical one.

As for algebra being a generalized form of number, that is about it. There is nothing mysterious at all. We express the rules that govern ordinary arithmetic in letters, so that the numbers can be non-specific. In an equation we ask that a particular expression should take a certain value, often zero.

Mostly the expression does not have that value and our graph shows the values it does take – but there are times when the equation balances. There may be several such points for a particular equation. A quadratic has two 'roots', a cubic three, and so on. Some are real numbers, some complex, but the rules are simple.

We now have the delightful example of how new ideas break through original assumptions. While we believe algebra simply to be about numbers, we are bound by the rules we stated. As we accepted the rules in Euclidean geometry, so in algebra did we stick to rules because that is the way numbers are. We shall begin to break through by playing what looks to be at first an easy game with enclosures.

Two tribes have certain land they claim. When they are at peace, everyone roams freely over the whole area of their two enclosures. This area is called, naturally enough, their UNION. When they fall out, they war over the small shaded area, called the INTERSECTION. As we see in the diagrams, we decide to use certain signs for these; we can choose any we like, but these are what we have chosen. We sometimes call them 'cup' and 'cap' because that is how we have drawn them.

Tribal boundaries

$\cup$ = union Shaded area = $A \cup B$

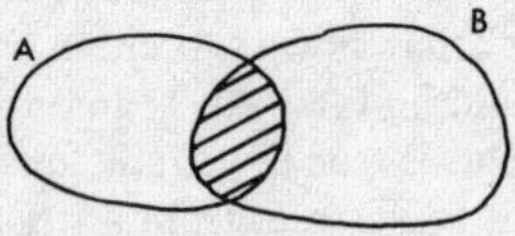

$\cap$ = intersection Shaded area = $A \cap B$

This is a simple enough definition. It has nothing to do with number, and seemingly nothing to do with algebra. But let us ask some rather obvious questions.

Is $A \cup B = B \cup A$? Of course, you only have to look at it.
Is $A \cap B = B \cap A$? Again, of course.

They seem trivial questions, yet suddenly it is looking like algebra, and the particular rules which govern it seem to be the issue again.

Bring in a third tribe, again with disputed areas. If all is well, does it matter if we join B's land to C's and then open up A's territory, or if alternatively we join up A and B first and then remove C's boundaries? A brief study of the diagram will assure you not. We show the expression $A \cup (B \cup C)$ built up from B, the unshaded region, combined with the rest of C, striped, and what remains of A, the hatched region. It gives the whole area. A little thought will show that so does $(A \cup B) \cup C$.

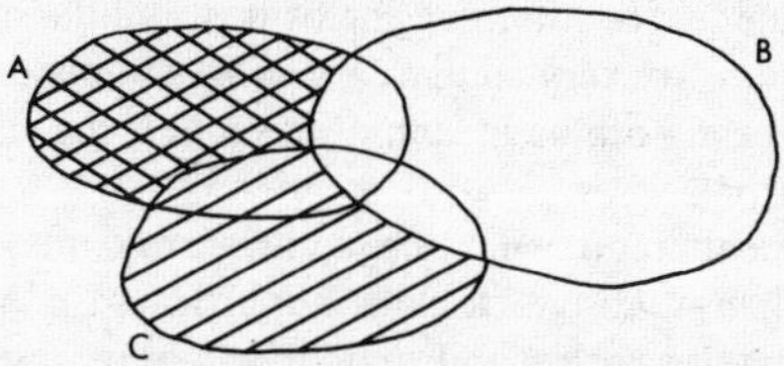

$$A \cup (B \cup C) = (A \cup B) \cup C$$

In the next diagram, B and C quarrel over the striped part, their disputed land; if A gets into it, which bit of that quarrel is he interested in? Clearly the innermost, double striped area. It is the same area that C would want to watch if A and B are quarrelling. These rather obvious facts are now stated in symbols:

$$A \cup (B \cup C) = (A \cup B) \cup C$$
$$A \cap (B \cap C) = (A \cap B) \cap C$$

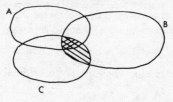

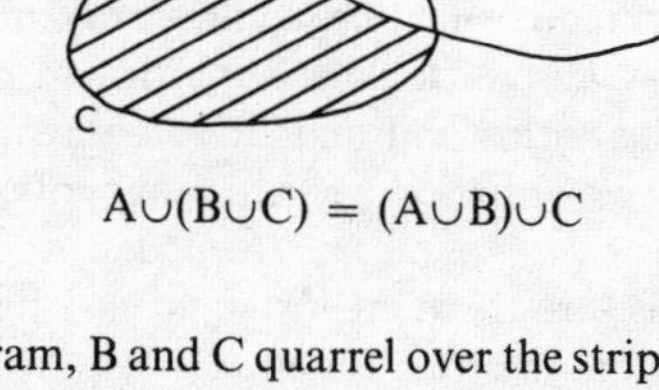

$$A \cap (B \cap C) = (A \cap B) \cap C$$

Finally, and we leave this for the reader, are these two statements true?

$$A \cup (B \cap C) = (A \cup B) \cap (A \cup C)$$
$$\text{and} \quad A \cap (B \cup C) = (A \cap B) \cup (A \cap C) \quad ?$$

Draw your own diagrams, and work out the areas we are talking about in each case, and see if both sides turn out the same for each line.

Suddenly, things are much more difficult than when we started. Yet there is a curiously familiar air about it all. The rules seem to be the same as those for numbers, though we are certainly not working in numbers. Let us put them side by side

$$\begin{array}{rcl@{\qquad}rcl} a+b &=& b+a & A\cup B &=& B\cup A \\ a+(b+c) &=& (a+b)+c & A\cup(B\cup C) &=& (A\cup B)\cup C \\ ab &=& ba & A\cap B &=& B\cap A \\ a(bc) &=& (ab)c & A\cap(B\cap C) &=& (A\cap B)\cap C \\ a(b+c) &=& ab+ac & A\cap(B\cup C) &=& (A\cap B)\cup(A\cap C) \end{array}$$

It seems that the rules for overlapping enclosures, and the rules for adding and multiplying numbers, are very similar. It is a most amazing result. The two processes seem to have nothing in common, yet at some very deep level there must be a connection.

It leads us to treat union as if it were addition, and intersection as if it were multiplication. When we discuss logic in our chapter on 'Modern Mathematics' we shall see that the worlds 'and' and 'or' can be represented as enclosures of the sort we just played with. So logic and number begin to show the same deep structure.

Samenesses of this sort are of great significance, but we must not ignore differences. There was another rule for enclosures:

$$A\cup(B\cap C) = (A\cup B)\cap(A\cup C)$$

The 'corresponding' rule in numbers would be

$$a+(bc) = (a+b)(a+c)$$

and that most certainly is not so.

The signs $\cup$ and $\cap$ have a symmetry with respect to one another that $+$ and $\times$ do not have.

There are other differences. If one enclosure happened to lie completely inside another (say B were inside A) then we get strange results such as

$$A\cup B = A \quad \text{and} \quad A\cap B = B$$

and we also get these results, which are not usual in algebra:

$$A\cup A = A \quad \text{and} \quad A\cap A = A$$

We said 'in algebra'. Perhaps we should revise that though, and say 'in number'. We are beginning to see that algebra could well have a wider meaning than generalized arithmetic. It could (and perhaps it is a matter for our choice) be used to describe any system of symbols operating with set rules.

In number we have a, b, c, d, . . . and the operations $+$ and $\times$
In enclosures we have A, B, C, . . . and the operations $\cup$ and $\cap$

We might well use the general terms 'elements' to cover numbers, enclosures, and any other strange phenomena we may meet. We shall need the term operations for processes between them.

We need some more examples to reinforce this widening of our notion of what algebra is about. Firstly let us take the moves of the four chess pieces, queen, rook, bishop and knight. There is no need to be a chess player to follow this work. We shall allow each piece in turn to stand in the middle of a 5 × 5 square, and we shall look at the squares it can 'cover', or move to.

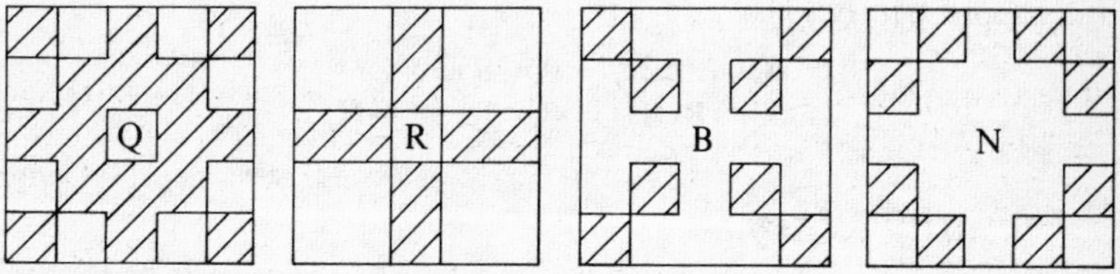

Queen Rook Bishop Knight

Although they are not exactly enclosures, we are going to use the union and intersection notion rather than addition and multiplication. First, let us make some observations on the patterns we see. Disregarding the square each piece stands on, we see that the Queen covers 16 of the remaining 24 squares, the other three each cover 8, all in different patterns. Let us call the whole square S and write down a few facts for discussion:

$$Q = R \cup B$$
$$Q \cup N = S$$
$$R \cap B = 0 \text{ (Zero)}$$

Think about each, then let us draw up a table for union and one for intersection

∪	Q	R	B	N
Q	Q	Q	Q	S
R	Q	R	Q	
B	Q	Q	B	
N	S			N

(1)

∩	Q	R	B	N
Q	Q	R	B	O
R	R	R	O	O
B	B	O	B	O
N	O	O	O	N

(2)

Check an item or two in the first table. If you take all the squares covered by the Q and then all those covered by the R you get . . . those covered by the Q. Obvious enough, because the Q can move like a R if she chooses. However, Q∪N = S. The moves of Q and N are complementary and cover the whole square. Notice that we have not filled in R∪N and B∪N. This is not because they do not cover certain squares altogether, but the coverage is not one of the four pieces, or even S or 0.

Our second table is better in this respect. It certainly has many zeros, but we get answers everywhere.

Are we still doing algebra? It is a matter of standpoint, but the mathematician would say we are. We cannot guarantee that quite the same rules apply. but try some:

Is $Q \cup (R \cup N) = (Q \cup R) \cup N$?
Is $Q \cap (R \cup N) = (Q \cap R) \cup (Q \cap N)$?

Identity elements

One of the central issues in mathematics is what does not cause change. In our chapter on geometry, we took special notice of what stayed unchanged. In algebra, let us look at what leaves something unchanged when we operate with it.

This is straightforward in arithmetic. If the operation we are concerned with is +, then the number that has no effect is 0.

$a+0 = a$, whatever a happens to be.

In multiplication, 1 is the important number:

$1 \times a = a$, whatever a is.

This sounds obvious, yet causes difficulty. At school the inability to handle zero and unity correctly often leads to trouble. Notice they have to be associated with a particular process. Adding 1 or multiplying by zero certainly does alter the situation. Yet people find it difficult always to remember that

$a \times 0 = 0$ whatever a is

There is always a tendency to think zero has no effect – and that is only true in addition.

How about identity elements in the enclosures? Suppose these warring tribes live on a large island which for them is the universe. We shall call it E (for Everywhere, if you like). We also have a tribe Z who, being bottom of the heap in everything, have no land. E and Z are going to play a special role, but each has to be put with the right operation. Remembering that an identity element will not affect the result when applied with a suitable operation, we have

$A \cap E = A$
$B \cap E = B$
$C \cap E = C$ and so on . . . but
$A \cup Z = A$
$B \cup Z = B$
$C \cup Z = C$

and we see that E and Z play roles like those of 1 and 0.

For Review

J M Dent & Sons Ltd
33 Welbeck Street London W1M 8LX 01 486 7233

We are pleased to send you a review copy of

The Everyman Reference Library

MATHEMATICS FOR EVERYMAN

by Laurie Buxton

Published: 18th April, 1985

Price: £3.95

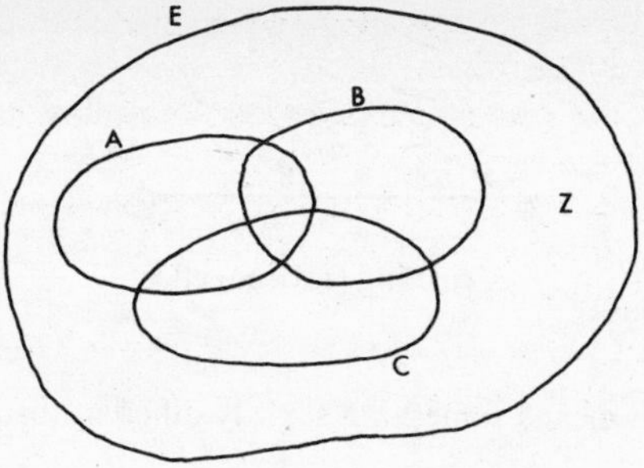

In our chess playing example, something about identity elements comes through in the table. Here S and 0 are our identities, as we shall point out with just two examples:

$$R \cup S = R$$
$$\text{and} \quad B \cup 0 = B$$

In our chapter on the calculus we shall take a special interest in a function that is not affected by the processes of calculus. Such elements matter everywhere throughout mathematics.

We shall make one last point about their importance in relation to numbers. Start with a number 'a' and look for the number that makes this happen:

$$a + x = 0$$

The number x is the 'inverse' of a. If for a we use the natural counting numbers, this equation is a formal way of saying what a negative number is. Looked at another way, it defines the inverse of + and that is −, and here we are talking of operations, not prefixes of numbers.

So far we have largely been in mathematics, despite some external references. We now look at a direct mechanical or engineering notion. When a force acts on something that is fixed at one point, and acts off-centre, it tends to turn. That is commonplace. In practice we are very often concerned with turning and with what torque we are applying.

In our diagram on p. 108 we have a starting point on the line of the force that we are going to apply. From there we draw a directional line (a vector) from the point we choose in the line of the force to the pivot of the object we are turning. Suppose this is $\vec{a}$ (the arrow indicates we have to consider direction; a by itself is simply the distance): The force is $\vec{b}$, again directional, again with b being the size of the force.

The turning effect is the product of these two:

$$\text{Turn} = \vec{a} \times \vec{b}$$

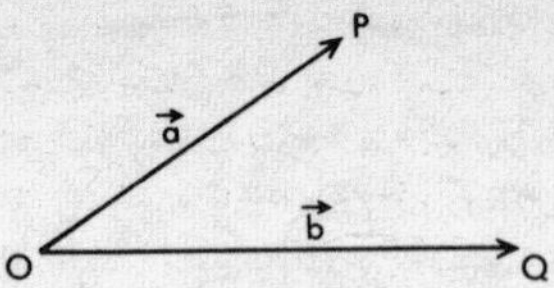

A turning (torque) effect

but multiplication means something slightly different, because of the directional issue.

Now if $\vec{b}$ were the position vector and $\vec{a}$ the force we would have a turn $\vec{b} \times \vec{a}$. Our natural belief is that it will not matter which is the positional vector and which the force. We might think

$$\vec{a} \times \vec{b} = \vec{b} \times \vec{a}$$

This would certainly fit in with everything to date. Indeed they are the same size, but just look at the diagram.

If $\vec{b}$ is the force turning about P it is anticlockwise. If $\vec{a}$ is the force turning around Q it is clockwise. They are opposite in sign. To make our algebra fit reality we must write

$$\vec{a} \times \vec{b} = -\vec{b} \times \vec{a}$$

That is a surprise, but it is thrust on us. Those vectors are not just numbers, so they do not behave exactly like numbers.

This is the fourth system we have worked on – numbers, enclosures, chess and vectors. Many of the rules are the same – but we cannot rely on it. Enclosures are not exactly like numbers. Nor are vectors. Vectors have a slightly different rule. For them, we no longer have a commutative law for multiplication. It may be a relief to know that

$$\vec{a} + \vec{b} = \vec{b} + \vec{a}$$

even if $$\vec{a} \times \vec{b} = -\vec{b} \times \vec{a}$$

In a relatively short time in the last page or two we have introduced the idea of a non-commutative algebra – simply because it made sense, and we took the view that the rules of any algebra are at our command. To put ourselves now in the position of mathematicians before those ideas is extremely difficult, yet before a possibility is seen, it takes a genius to resolve it. The first introduction of non-commutativity came with William Rowan Hamilton and his development of quaternions (a notion related to vectors). It was after fifteen years of intense intellectual effort that he considered the idea that an algebra could be consistent even with a non-commutative law.

Hamilton's work was enormously extended by Grasseman, who created many algebras, including the basis of the tensor calculus which was essential to Einstein's theory of general relativity. As we shall see with the conic sections in 'Mathematics in Action', there are various examples of mathematics being created entirely for its own sake, the purest of the pure, and then finding the most direct of practical uses. Grasseman had no notion of how any of his algebras might fit a real model; perhaps he did not care. Yet the work was ready to hand when Einstein needed it.

Let us see where we stand with algebra – or rather algebras. We have some elements; we do not mind what they are, they can be numbers, enclosures, chess moves, vectors, what you will. Or you call them just elements and leave it at that. There are some operations; as few or many as you like. They may resemble $+$ and $\times$ or may be very different. There are then some rules that govern relationships between the elements and the operations. They are at our choice, with one large proviso. Whatever results we arrive at, using our rules, we must never produce a contradiction. The essential feature of mathematics is that it must be internally consistent. That is its only real test. Any piece of it may or may not fit some situation in the real world. If it does, it generally proves remarkably useful. If it does not, it does not mean that it is not good mathematics.

So if you like to list some elements – a finite number or an infinite number, and some operations, say * and $\overline{+}$ (though we could do with one, or have more than two if we wished) – and we then write down some interesting rules, we are in business. We have an algebra.

8

The calculus

The word calculus comes from the Greek for a pebble, or a stone used in calculation, but the naming is neither felicitous nor helpful. The public image of the calculus is of something decidedly advanced in mathematics, and rather dignified. It is often referred to as 'the' calculus, while we do not talk of 'the' arithmetic or 'the' trigonometry. The definite article lends a certain touch of class.

There are some difficult ideas at the very beginning of the calculus, and if we get too deeply involved with infinitesimals we can be led out of our depth. Some of the notions are not easy to square with our basic intuitions, and this adjustment is necessary before we can feel we understand something. In our next chapter on probability we show how easy it is for intuitions and mathematics to be at variance. Yet the calculus is designed to deal with everyday problems which we regularly meet. We do not have to use it to cope, but the understanding of them lies through the calculus. The main issues are in movement, in words such as speed and acceleration. In an increasingly technological society we are constantly told of new and expensive toys with speeds of 115 mph and acceleration 0–60 mph in 9 secs. Our courts constantly discuss such matters in the flood of traffic offences with which they deal. That is where we shall start.

A stone falls. Through the centuries we have seen things fall when we drop them, and we became so used to it that it seemed pointless to remark upon it. It is simply the way things are. Perhaps there is something in the 'nature' of a stone that makes it fall; perhaps it is part of some god's design. 'Explanations' of this sort satisfied many. Many animals other than ourselves solve the problems they meet, often through a need to survive. We may be the only animal that asks the question 'Why?' In our historical development the 'why?' of the falling stone (or some say apple) had to wait for its first scientific explanation till Newton, but powerful minds before him had considered the problem. Aristotle held that how fast an object fell depended on how heavy it was and that in fact it was proportional to its weight. An object ten times the weight of another fell ten times as fast. This was not only untrue, it did not begin to discuss what was meant by 'fast'. Galileo, one the most imaginative and creative minds of all time, but whose methods had an engaging simplicity, is said to have taken two weights, one much heavier than the other, and dropped them from the leaning tower of Pisa. They hit the ground at the same time. We have, therefore, known this fact for some four hundred years. In our society today, probably most people know it, but there must remain a substantial number who do not. It may be that as they go about their ordinary business it does not matter to them. Sherlock Holmes remarked once on his complete lack of interest in whether the earth went round the sun or the sun round the earth. Had Galileo not expressed certain views on this matter, he might have escaped an uncomfortable interview with the Inquisition. If and why we should know such things is an issue we shall consider at the end of this book. Let us continue with our falling stone.

Galileo was not content with the information that all objects fell at the same rate. After sorting out the reasons for feathers floating he wanted to know exactly how a stone fell. The move from the qualitative to the quantitative is a significant one. Science is about prediction, and the text of a scientific prediction lies in experiment. Galileo wanted to know where the stone was at any time, how fast it was going and how it was accelerating. It is difficult to put ourselves back in time and see the power of mind that was needed even to pose these questions. Posing questions is at a higher level of sophistication even than solving them. They are questions which only a tiny proportion of the population can solve today. Ask yourselves the questions. Drop a stone from a high window. Have you any idea how much further it drops in the second second than the first? If the ground is 100 ft below, how fast will the stone be travelling? There are many such questions, and few people could answer them even nowadays. They matter. If you are designing a safety helmet for people on a building site, it matters how fast that brick is going. If a car crashes, the deceleration that a human body can tolerate is crucial. It may be that the man in the street does not need to know, but the people who design the things he uses need to know.

Galileo grappled with and solved the problems. He did not have instruments accurate enough to measure the facts about a body falling under gravity (we didn't have gravity then, that came with Newton!) so he allowed them to slide or roll down inclined planes. This reduced the headlong rush to manageable proportions for him. He learnt what he wanted to know, and began to gather together facts about distance, time, speed and acceleration, that would lead to the calculus a century later.

Newton remarked that if he had seen further than others it was because he had stood on the shoulders of giants. Galileo was one of the giants. Yet in any contest for the most powerful mind of all time, Newton would be the man to beat. His extraordinary capacity lay in accepting the vast amount of material that had accumulated through observation and reducing it, through asking 'Why', to simple rules from which all those observations could be deduced. He did this both with the material that Galileo had produced, and the very extensive observations on planetary motion by Kepler.

In effect he said that forces were responsible for motion; a force is something we all have direct experience of through the muscles of our own bodies. He said that a constant, steady force produced a constant acceleration. That was the surprise. Ask the ordinary person even nowadays what a constant force would produce and he will think it to be a constant speed. We are aware that we need a force to achieve acceleration, and experience this directly as we put our foot down and pull away in a car.

A result of this is that if there is no force then the object either stays still or moves at a steady speed. This does not match with our experience. When we take our foot off the accelerator we will eventually slow right down, even without braking. This is because the forces are not only those produced by the engine, but the external frictional and air resistance forces. A stone skimmed across a slippery icy pond is nearer the statement made by Newton. Much more importantly, in space travel, once we get out there, we switch off and move on with undiminished speed through space.

The force exerted by the earth on any object we drop is such as to give it a fixed acceleration of 32 ft/sec/sec (or 9·8 m/sec/sec). We need to examine these units and see if they make sense to us. It is the second 'sec' that is the problem. The meaning is simply this. Acceleration tells you how quickly something is gathering speed. Gravity at 32 ft/sec/sec increases the speed from say rest to 32 ft/sec after 1 sec, to 64 ft/sec after 2 secs, to 96ft/sec after 3 secs. Compare this with the advertisement about the car. It told us how quickly the car increased its speed to 60 mph. If we change that speed to ft/sec it comes out to 88 ft/sec. So a falling stone is going at well over 60 mph in 3 secs of fall (slightly diminished by air resistance).

Look at this 3 sec period and let us work out various speeds, times, and distances, as Galileo did. In the first second we start at rest, gather speed uniformly and reach 32 ft/sec. On average we are doing 16 ft/sec and in that second will fall 16ft. (We must be careful about taking averages in this way. The uniform acceleration permits us to here.) So if we fall from a second floor balcony we will hit the ground at about 22 mph, which will do us a lot or a little damage depending on the ground.

Diagrams can help a lot. Our graph on p. 113 shows the speed against the time. It goes up steadily (the acceleration is constant) over the three seconds, and 'steadily' means it is a straight line. We have recorded our speeds of 32 ft/sec, 64 ft/sec and 96 ft/sec at successive seconds, but for the moment are considering just the first second. We assumed that in that second we would cover the same distance if we travelled at 16ft/sec throughout. Our dotted line and the shaded area under it represent this fixed speed and the distance. (The area of the rectangle is 16ft/sec × 1 sec and that is 16ft.) Effectively we are saying that the area represents distance, and instead of the rectangle we could have taken the triangle OAP, which the eye can see has the same area. (Move the lower shaded triangle on to the small unshaded one.)

In the second second we start at 32 ft/sec and end up at 64 ft/sec. We can again say that this covers the same distance as a steady 48 ft/sec, that is 48 ft, or

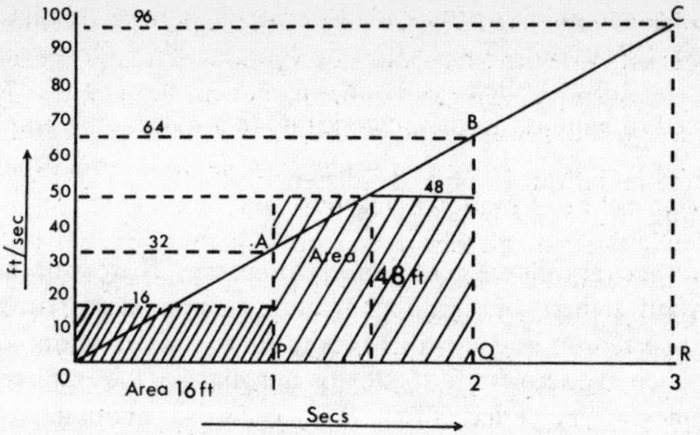

Falling under gravity

we could take the area directly of PABQ. This would not look very different, for the rule for our area like this is the average of the parallel sides (48ft/sec) times the distance apart (1 sec.) – so we are back with the 48.

Using what we have gained, let us look at the total distance fallen in 3 secs. It will be the area of the big triangle, OCR. The formula for a triangle is $\frac{1}{2}$ base × height and that is $\frac{1}{2} \times 3 \times 96$ or 144 ft. That is a long way to fall in three seconds, and as we said earlier the speed is about 65 mph.

In these calculations we ignore air resistance. We had better discuss the feather. It certainly does not follow this law. It does have a pull on it which would make it accelerate at 32 ft/sec/sec, but the air resistance matters with a light body with a large area. Put the feather and the stone in a vacuum jar and they fall together. The first men on the moon also demonstrated this in the airlessness of the moon's surface. Even our stone begins to be affected as its speed increases and eventually the air resistance balances the pull of gravity, but only when the stone is moving very fast. A human body reaches a maximum speed of about 120 mph. It is not sensible to hit hard ground at this speed, but people have survived falls from aircraft with no parachute, by falling through the branches (not too thick) of a tree, or into a deep snow drift. The necessary conditions for survival are not such that one should experiment.

Galileo had sorted out what we have just done, but there are some issues in it we need to pick out a bit more clearly before we get to the calculus. We concluded that the area under the line represented the distance travelled; where does the acceleration appear? We said that steady increase in speed meant constant acceleration, and that accounted for the line being straight. Look at its slope. The speed rises by 96 ft/sec in these 3 secs. This means that in every second it rises 32 ft/sec. So for acceleration we have the rather unusual-looking units – ft/sec/sec. This acceleration of 32 ft/sec/sec is the acceleration due to gravity, often called 'g'. We did not need to take the whole line; any part tells us our slope. 64 ft/sec in 2 secs is still 'g'. Or if we move from A to B we still go up 32 ft/sec in 1 sec.

We need to hang on to these two ideas. In a speed–time graph:

(1) The area under the graph is the distance.
(2) The slope of the line is the speed.

Look at our next graph, the general speed–time one. There is nothing steady about the motion there, but it is a much more common occurrence for us for the speed to vary up and down. Motion under gravity is one of the few situations where the acceleration is steady. Imagine that the graph represents a car's movement over a period of time. What can we say about it? Certainly the general trend overall is an increase of speed, or an acceleration, yet it is not the same all the time. The last section of the graph is more like the first graph, a steady increase of speed, yet earlier, say between the two vertical lines, the speed had dropped slightly. Typically it might be a car trying to pull away with a fair amount of traffic around. Most of the first section is fairly flat; the car is stuck in a steady stream. It then breaks away, builds up speed, but is then forced to slow, owing to more traffic. Finally it is in the clear, building up towards its maximum speed.

We have put no units on our two axes, yet we have managed to interpret what has been happening. Inside the car the driver too has information as to what is happening. He has a speedometer, a dial that records distance, and he knows how far his foot is down on the accelerator. It is perhaps odd that with

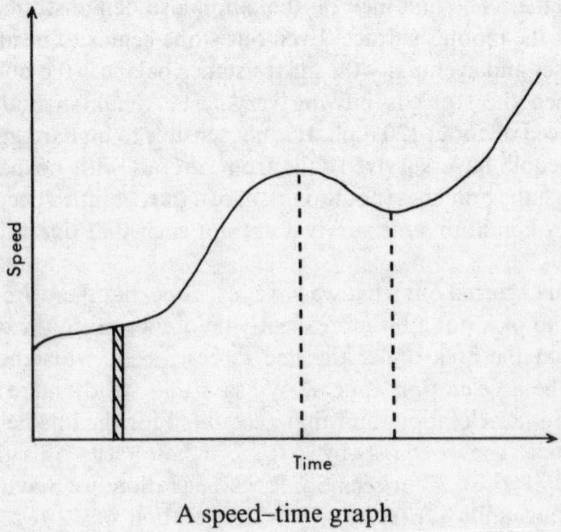

A speed–time graph

all the dials we have in a car we do not have one for acceleration. Can we get similar information from the graph?

At any time the height of the line from the time axis up to the curve is the speed. Imagine the graph turned on its side, and suppose that the car's speedometer is of the linear type, like the shaded bar. As it moves out and back, the end of it traces exactly the curve we have. As before, the area under the graph is how far the car has travelled. We are not proving this here, merely asking you to extrapolate from the falling stone example, though shortly we shall get nearer a proof.

What about the acceleration? Previously we measured the slope of the line, or its steepness. We must attempt the same. We said that the first part was fairly flat, and therefore there was little acceleration; we said it then increased, then went negative (we slowed down) and then was quite steep and continuous. The acceleration is measured by the steepness or the slope of the curve.

Through our discussion of movement, we now believe that in a graph illustrating it: .

(1) The acceleration is given by the slope of the curve.
(2) The area under the curve is the distance travelled.

We have reached the calculus. We can say that the differential calculus applies itself to the problem of finding the slope of a curve, and that the integral calculus deals with the area under the curve. We came at it through motion and shall return to the motion later. For now, let us try to find the slope and area under one of the simplest curves we know. In 'Back to Basics' we looked at the square numbers, 1, 4, 9, 16, . . . and in 'Mathematics in Action' we shall draw a parabola based on those numbers. That is the curve we shall try to work with (see the graph on p. 116).

As yet we do not know exactly what we mean by the slope at a particular point on this curve, but we could find the average slope of chunk of it, say from A to B or from B to C.

To go from A to B is 1 unit across and 5 up. We call the slope 5/1 or 5. (Notice that three squares that way is only one unit – but we could have used the small square as a unit that way as well, provided we are consistent.) From B to C the slope is 7/1 or 7. The gaps between squares, you may remember, were the successive odd numbers. The move from 5 to 7 shows the slope is increasing, as we expect. The important thing is to measure, as Galileo first showed us.

What is the slope at B? That is the real question. We know that if we were going up in straight line chunks, the line leading up to B would have slope 5 and the line going up from it would have slope 7. But the graph is a curve, not a series of straight lines. If we imagine the curve as a hard ridge and push a ruler up to it, that ruler has a slope, and if it rests against B then it measures the slope at B. So there is a practical method of doing it. The calculus aims at calculating

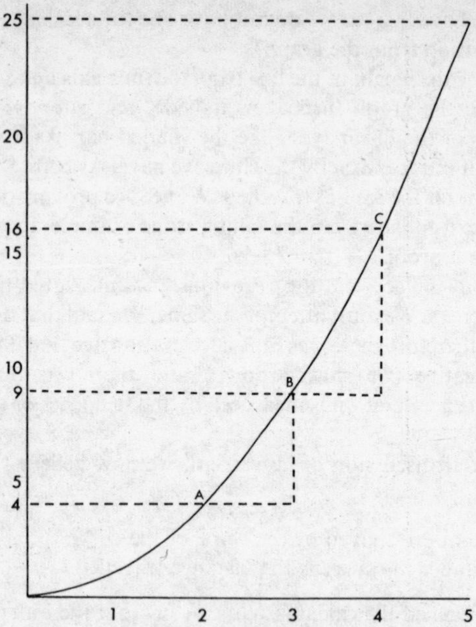

The slope of $y = x^2$

that slope. Practically there is no difficulty, other than getting the line to touch exactly at B. This line is called a tangent, because it touches (Latin: *tangere*, to touch). We notice that it slopes more than AB and less than BC, which is no surprise, but we need to note it.

At this stage we know the slope lies between 5 and 7. Now 'blow up' the picture near the point B. Our next diagram plots $(2{\cdot}9)^2$ or 8·41; 9 and $(3{\cdot}1)^2$ or 9·61. This time we have joined these three points by two lines. Their slopes are so close that the bend is barely perceptible. In fact the lower line has a slope of ·59/·1 or 5·9, and the upper of ·61/·1 or 6·1.

We know that the tangent still lies between them (we could not draw it distinguishably on the paper) so its slope is between 5·9 and 6·1. We can then blow up the picture another ten times, and we would find the slope of the tangent pinned even more closely. It is the process of taking smaller and smaller gaps around the point that led to the term 'infinitesimal' calculus. With a micro computer at your elbow it is a small matter to write a program to find the slopes just above and just below that of the tangent, with the gaps on each side of 3 getting smaller and smaller. The numbers would drive one to believe the slope at that point is in fact 6·0.

There is a lingering doubt about such 'limiting' processes, and it takes a

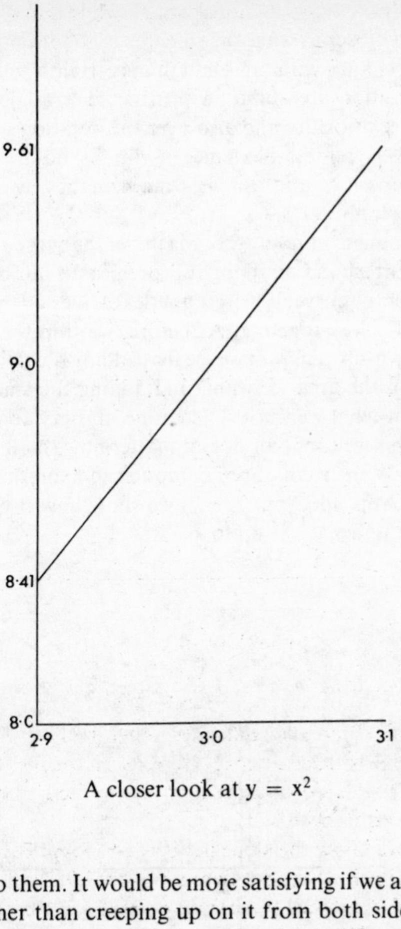

A closer look at $y = x^2$

while to get used to them. It would be more satisfying if we actually calculated it at the point rather than creeping up on it from both sides. This view was shared by mathematicians who felt a great need to justify and put the method on a firm footing, but the analysis needed is somewhat intricate. Here we seek to convince rather than prove.

For this curve where $y=x^2$ (every vertical column to the graph is the square of the number at the bottom) the result, wherever we try, comes out to be 2x. So the slope at 3 is 6, as we saw, the slope at 1 is 2, and the slope at 15 is 30.

From here the differential calculus seeks to find the slope of all sorts of curves, which may or may not have something to do with motion. The methods may be varied, and the theory extensive, but we are always tackling the same problem – what is the slope of the tangent at a point?

Newton and Leibniz invented the calculus independently. Newton called his approach the 'method of fluxions' which, while it may suggest a stomach disorder, is in fact closer to what the calculus is about than the word we now use. 'Flux' meant change with time, and this is certainly what the problems on speed and acceleration are about. A protracted wrangle, of little credit to anyone, arose over priorities and also over the notation used in dealing with problems. This was regrettable, since as far as notation was concerned, Leibniz's was superior, and English mathematics was inhibited by its adherence to Newton's.

The other problem – finding the area under the curve – is approached in a way with some corresponding thought processes. In our diagram we wish to find the area under the curve. We draw a series of lines up to the curve from the x axis, moderately close together. As before, we trap the area between two limits. In the 'blown-up' diagram we see that adding the tall rectangles gives an answer more than the area we want, and adding the smaller ones gives an answer a bit below what we want. There is no mystery about it, if we want to calculate these sizes we can. Nor does it matter how small we make the width of the rectangles. With calculator or computer to hand the task is not what it once was. In the Appendix (pp. 236–7) we show how it works out when we calculate the area under $y=x^2$ up to $x=3$.

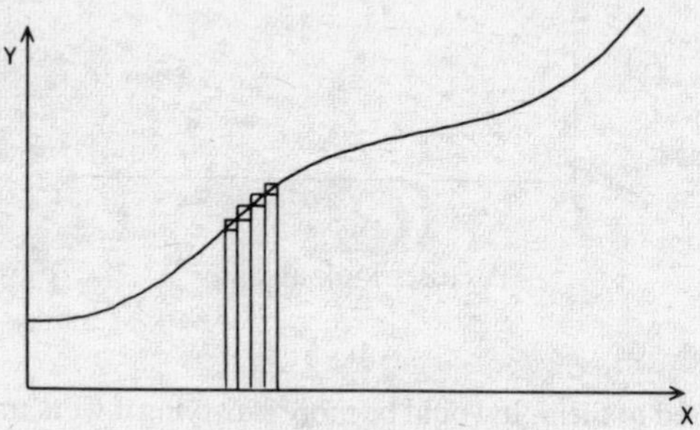

The area under a curve

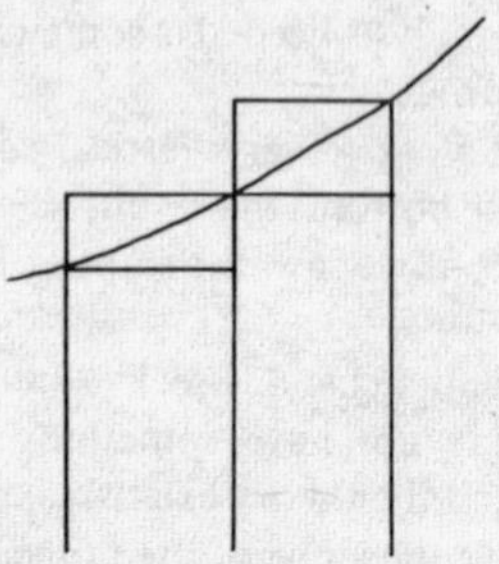

A closer look at the area under a curve

There had been a near approach to integration many centuries ago by Archimedes, who is rated by many as in the top three, mathematically, with Newton and Gauss. In working on the area of the circle he calculated the areas of regular polygons inside and outside a circle, knowing that his answer lay between them.

We are now possessed of two different operations, differentiation and integration, which we can apply to expressions in x (or in any other letter we choose). A great deal of effort by the mathematical community worked out, in the ways suggested, the differentials (sometimes called derivatives) of a large range of expressions. These could be stored as information (rather as we must at first have learnt and stored our multiplication tables) to be used when we wish. The process of finding them is lengthy and tedious and we would not wish always to repeat it.

So we know that the derivative of x^2 is $2x$, of x^5 is $5x^4$, of x^9 is $9x^8$ and so on. We know the derivative of sin x is cos x and of cos x is $-\sin x$

We know that the integral of x^2 is $x^3/3$ and of x^5 is $x^6/6$ and so on. The integral of sin x is $-\cos x$ and the integral of cos x is sin x.

The collection of pieces of information leads us to a very important discovery. Integration and differentiation are inverse processes. If you integrate an expression and then differentiate it, you are back where you started. This fact is known as the fundamental theorem of the calculus.

We now begin to see some resemblances between these two operations and the 'four rules'. Adding, multiplying, dividing and subtracting are operations on numbers, integration and differentiation on expressions or 'functions'. And as adding and subtracting are inverses (for if you add something and then subtract it you are back where you started) so are multiplying and dividing and differentiating and integrating. Again, a basic theme recurs as we advance our mathematical knowledge. The analogy is not precise. The four rules are known as binary operations (not the same as binary numbers) because they produce a result between two elements. Our calculus processes are operations on a single function.

The 'stay as you are' idea is curiously important in mathematics. It may seem the least sensible idea to consider, but this is not so. In our number system zero and unity play a particularly important role, and part of the importance lies in the fact that adding zero has no effect and that multiplying by one also has no effect. Within the calculus there is an equally important function, which is not affected by differentiation and (effectively) not by integration.

This function is so important that we must look at it, even though it means a little algebra and an approach that might not appeal to the purist mathematician.

Take this string of expressions involving x:

$$1+\frac{x}{1!}+\frac{x^2}{2!}+\frac{x^3}{3!}+\frac{x^4}{4!}+\ldots$$

Immediately we have a problem with another symbol. The exclamation mark after a number means that we want to multiply it by all the numbers below it. We could have written the series as

$$1+\frac{x}{1}+\frac{x^2}{2\times 1}+\frac{x^3}{3\times 2\times 1}+\frac{x^4}{4\times 3\times 2\times 1}+\ldots$$

except that it becomes a little clumsy by the time we reach a dozen or so. We now propose to differentiate this, term by term. Knowing what we do about differentiating powers of x, we have no problem after the first two, but we should look at these. The graph of y = 1 is a straight line going horizontally, for it means that y is 1 whatever the value of x, and that is what the line expresses. Its slope is zero; it is flat. So when we differentiate 1 (or any number for that matter) we get zero. The next graph shows y = x, where the two values march hand in hand. One step along, one step up, and the slope is 1.

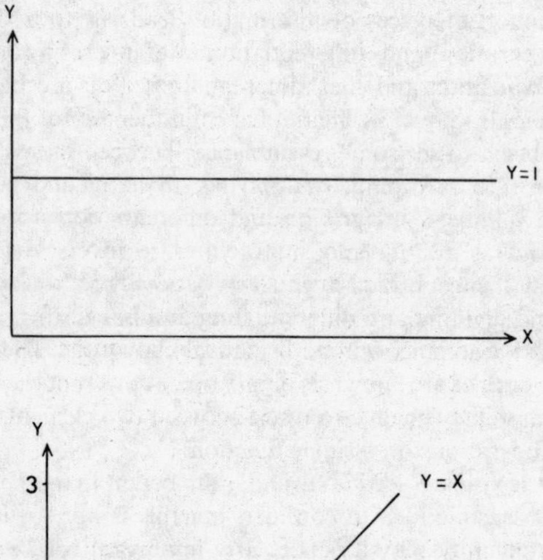

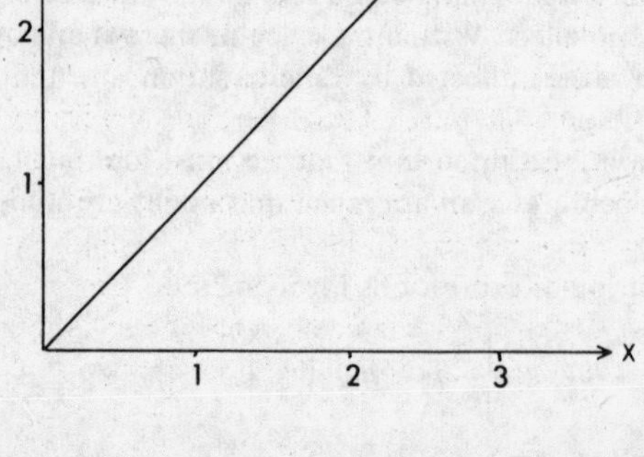

Now to differentiate, term by term:

$$0+1+\frac{2x}{2\times 1}+\frac{3x^2}{3\times 2\times 1}+\frac{4x^3}{4\times 3\times 2\times 1}\ldots$$

or, cancelling:

$$1+\frac{x}{1}+\frac{x^2}{2\times 1}+\frac{x^3}{3\times 2\times 1}\ldots$$

and we are back to our original series. 'What happens at the end of the series?' you say. 'Well,' says the mathematician, 'we will let it go on for ever, and then there won't be an end.' You may think it is cheating; if so, we discuss it further in the Appendix p. 237.

The series, then, is unchanged by differentiation. Let us look at the series with x put equal to 1. We get

$$1+1+\tfrac{1}{2}+\tfrac{1}{6}+\tfrac{1}{24}+\tfrac{1}{120}+\tfrac{1}{720}\cdots$$

where we have worked out the bottom of each fraction. Our calculator (fortunately equipped with reciprocals, or 1 over something) finds this to be 2·7180556. It is obvious that the bits we add on are rapidly getting smaller and even if there are a lot of them (like an infinite number!) it will not bump it up too much. In fact, with $x=1$, the expression is 2·7182818 . . . .

This number is called e. It is important, as we said, in somewhat the same way that 0 and 1 are important. In fact, together with π (the ratio of circumference to diameter in a circle) these four numbers are the really significant points on our number line.

If we were to put $x=2$ in the series and again resort to our calculator or a micro we would get, when we had added enough terms, the number 7·3890561. This turns out to be the square of e. So putting in 2 squared it. It suggests the x should be a power of e.

Following this through we now write:

$$e^x = 1+\frac{x}{1!}+\frac{x^2}{2!}+\frac{x^3}{3!}\ldots$$

and we list in our stock of results that the differential of e^x is e^x.

We can now clear up the point about integrating it. If we integrate e^x we get e^x but there could be a number added to it, since when we go the other way e^x+2 (say) also differentiates to e^x. Here the details are not so important; the real issue is the stability of e^x under attack by either a differential or an integral.

We now have a diversion for the more mathematically minded. For others, at least marvel at the result when we get there. In the expression for e^x we can put in any value of x that we like. For reasons that seem mysterious, but whose purpose we shall shortly see, we put $x=\pi i$. The ratio of the circumference of a circle to its diameter is reasonably familiar. The number i, $\sqrt{-1}$, is discussed in the chapter on number. We do not need to understand it here. All we need is to recognize that if $i = \sqrt{-1}$ then $i^2 = -1$, which after all is what we mean by square root. It follows that $i^3 = (-1)i$ or $-i$ and $i^4 = 1$. For higher powers we go around the four values i, -1, $-i$, and $+1$, and using these we get:

$$e^{\pi i} = 1 + \frac{\pi i}{1!} + \frac{\pi^2 i^2}{2!} + \frac{\pi^3 i^3}{3!} + \frac{\pi^4 i^4}{4!} + \ldots$$

$$= 1 + \frac{\pi i}{1!} - \frac{\pi^2}{2!} - \frac{i\pi^3}{3!} + \frac{\pi^4}{4!} + \ldots$$

It will now help if we separate out the terms with i in them and the others.

$$e = \left(1 - \frac{\pi^2}{2!} + \frac{\pi^4}{4!} - \frac{\pi^6}{6!} + \ldots\right) + i\left(\frac{\pi}{1!} - \frac{\pi^3}{3!} + \frac{\pi^5}{5!} - \ldots\right)$$

At least both series are nicely patterned. It would be a great task to calculate these series, partly because the expressions involve π and partly because each goes on for ever. We must have a quiet word with the micro. The computer is addressed thus:

```
10  LET T = 0
20  LET K = 1
30  FOR N = 1 TO 9 STEP 2
40  LET K = K × π² x (−1)/n(n+1)
50  LET T = T+K
60  NEXT N
70  PRINT T
```

This asks it to add up the terms in the first bracket (not counting the 1), and only as far as the fifth of these terms. It instantly responds $-2{\cdot}0018291$. We step up the 9 in stages of 2 until it gets to 17. This means we are adding nine terms together. At this stage it reports (-2) and continues to do so for all greater numbers of terms. That does not mean that it is exactly (-2) but that the computer (working to 8 decimal places) cannot tell the difference.

A similar process on the other series, and again after 9 terms the micro reports the answer 0 and sticks to that.

We are in another limiting process. We keep adding terms for ever but they have become so small they have no perceptible effect. Our acquaintance with limiting processes should now assure us that $e^{\pi i} = 1 - 2 + 0$.

If not, we can only say 'Oh ye of little faith!' Sorting out this equation we get (and we shall print it large)

$$e^{\pi i} + 1 = 0$$

The linking of e, π, i, 1 and 0 in this manner is a truly moving experience!

There are more elegant, more correct, more mathematical ways of reaching this result, but the pure number-crunching power of even a little micro has tempted us into this path. Where once we dealt in infinitesimals, or worried about the tail of a series, we now can at least see what is happening even if it does not amount to a proof, by setting the computer at it.

At this stage we seem once again to have departed from the real world. We started in practical situations involving movement and went into graphs and pictures and now we have an odd sort of number with rather mystic properties. But the route back to the real world is immediate. Many real life problems come out in terms of the 'exponential function', as e^x is called.

Look at the graph of e^x. It rapidly becomes very steep, more quickly than any expression involving powers of x, and is a picture of growth and decay that matches in importance the 'normal distribution' curve we meet in the chapter on probability.

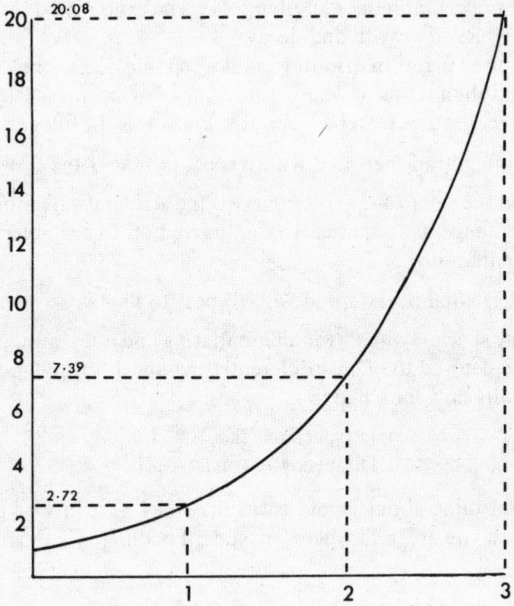

$y = e^x$

Invest your money and leave it in at constant interest. If it is worked out once a year and added to the capital, the amount begins to grow more and more quickly. The more there is the more it grows. Compounded once a year money goes up in jumps. The curve of growth of money gets smoother if it is compounded say every month. As we add the interest at closer and closer intervals, so do we get nearer and nearer the exponential curve. The exponential grows at a rate proportional to how much is there. (This is something like saying its derivative equals itself.)

Buy a new car and you are on the same curve, but the other way. Think of the mirror image. Its value starts near the top and drops alarmingly. As the years go by it gets flatter in price and the depreciation each year is less in money terms, even if the same in percentage terms.

Some moulds, or yeast, grow for a while exponentially. There was just the possibility with the first atomic explosion that it would grow exponentially and 'take off' – and us with it. Radioactive 'decay' means that the rate at which a radioactive element is changing its form is exponential – again we are coming down the slope.

Whenever the change depends on what is there the form will be exponential. The mathematics does not mind whether what is there is invested money, a car price, uranium or anything else. The exponential function lies at the heart of issues of growth and decay.

We promised to return to motion, and we should do so. Look again at our speed-time graph in terms of what we now know of calculus. The area under the curve is the distance travelled. We also know it is the integral. So we say:

'The integral of the speed with respect to time is the distance.'

We know that the slope of the speed curve is the acceleration (although we did not draw any tangents to that curve, but moved off to a simpler one). Slope means differentiation, so:

'The differential of the speed with respect to time is acceleration.'

There are special symbols for differentiating and integrating, but we shall make life simpler and use D for differentiating and I for integrating.

Our results now look like this:

I (speed) = (acc) . . . I
D (speed) = (dist) . . . II

The important thing about D and I was that they were inverse processes. If you apply both you are back where you start. Try differentiating both sides of equation I:

DI (speed) = D (acc)

and since DI means 'stay as you are' then we get

(speed) = D (acc)

We now rewrite the two equations, switching our last result round:

D (acc) = (speed) . . . I
D (speed) = (dist) . . . II

This is rather pretty. It shows that acceleration has exactly the same relation to speed as speed has to distance. How many motorists, accustomed as they are to the quantities we are dealing with, could formulate that connection – not the calculus connection involving differentiating, but simply the idea that the two connections are the same?

Back now to the leaning tower of Pisa. We drop our weight from rest at the top. The acceleration due to gravity is fixed. (We will use 32 ft/sec/sec, though we doubt if Galileo did.)

$$\begin{aligned} \text{acc} &= 32 \\ \text{speed} = \text{I(acc)} &= 32\text{t} \\ \text{dist} + \text{I(speed)} &= 16\text{t}^2 \end{aligned}$$

(t, the time, is the variable now: we treat it exactly as we treated x)

So we know immediately its speed is 32t and the distance fallen $16t^2$. This means that after 3 secs it has fallen 144 ft and is travelling at 96 ft/sec. We can also do reverse tricks. If we want to know how long it takes to fall 400 ft we fill in the distance and get

$$\begin{aligned} 400 &= 16\text{t}^2 \\ \text{or} \quad 25 &= \text{t}^2 \end{aligned}$$

and so it has been falling 5 secs.

The ideas behind the calculus need intellectual grasp. Working with it can often be as routine as multiplication.

Calculus therefore deals with motion, growth and decay. That is a substantial amount, but there are more issues it can tackle. Another important area is the determination of maxima or minima. We shall start with a simple problem, and when we have sorted that out, indicate much more complicated areas where the calculus is of value.

We are given a length of fencing and told we may enclose as much land as we can of a rectangular shape against a wall. Our diagram shows the situation.

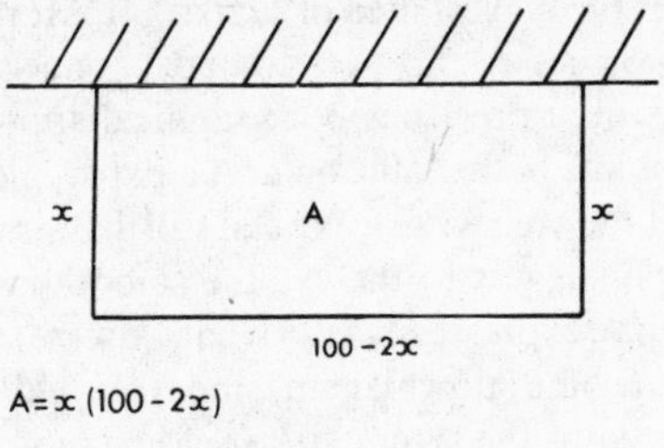

The allotment area

If they gave us 100 ft of fence, how long would we make x to enclose as much as possible? This again is an interesting problem to test your intuitions on, but not one that you are likely to find surprising.

There are various methods of attack, and they illustrate the variety which mathematics offers. We need a bit of algebra. If we put it out a distance x, then with the two ends being x, it leaves $100-2x$ for the other side. Taking the rectangular area as length times breadth we have our formula:

$$A = x(100-2x)$$
$$\text{(or } 100x - 2x^2 \text{ if we like)}$$

But if algebra is not your cup of tea, read on; we do not use it all the time.

The table of values puts x at various amounts and works out the area you get. Interestingly, you do not need to use the formula if you do not want to. If you are making a long thin rectangle with $x=5$ it needs no algebra to know that the long side is 90 and the area therefore 450.

x	5	10	15	20	25	30	35	40
A	450	800	1050	1200	1250	1200	1050	800

The allotment area: table of values

Our table could therefore represent the trial and error approach to mathematics, and this is not to be sneered at. It can lead to an answer, but even if it does not, valuable insight into what is happening can be gained. Just trying a string of numbers, every 5 just to get a spread, is a sensible start. When we have done it, pattern appears. As we increase the length of the piece coming out from the wall so the area increases, but when we get to 25 it then starts down again, and starts hitting the numbers we had on the way up. Simply from trying these numbers, it becomes quite clear that making x 25 ft long and the long side 50 ft gives us the biggest area of 1250 sq. ft. Moreover it is the sort of answer we might have guessed. So trial and error gets a tick.

Perhaps slightly more precise is our graphical approach on p. 127. We merely enter up the points in the table on graph paper, plotting the successive values every 5 against the area we get. We can still do this without algebra. We get eight points which suggest to the eye the smooth curve we have drawn through them. The curve expresses exactly what we were saying. The area goes up to a peak and then comes down again, smoothly and symmetrically. Had we plotted one more point the symmetry would be more complete.

From the curve it is even more clear than from the table where the area 'peaks'. The visual impact of graphical representation such as this is very powerful. The way the calculus tackles this problem hinges upon the appearance of the graph. The graph slopes less and less steeply as we go up the left hand side, and then slopes downhill (negative slope) as we pass the peak.

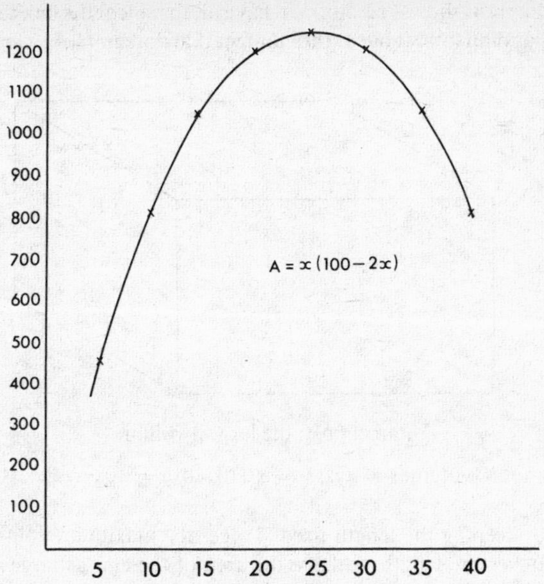

The problem in graph form

At the peak, a tangent would be flat. The slope is momentarily zero at the top point. It is this fact that the calculus uses.

We have

$$A = 100x - 2x^2$$

and we need to find where the derivative is zero (flat tangent):

$$DA = 100 - 4x$$

and this is zero when $100 - 4x = 0$ or $x = 25$.

Thus in a few lines the answer is calculated. The principle depends on the picture, but the method is precise. What is more important is that it can be applied to examples where trial and error on graphical methods is not effective.

Another example will make this point:

'The regulations for posting boxes say that for a rectangular box the sum of the length and the girth must not exceed 10 ft. What is the largest volume you can send by post?'

The diagram shows the box. We have all three lengths unknown. We do know they add up to 10 ft but at the moment that does not help a lot. First look

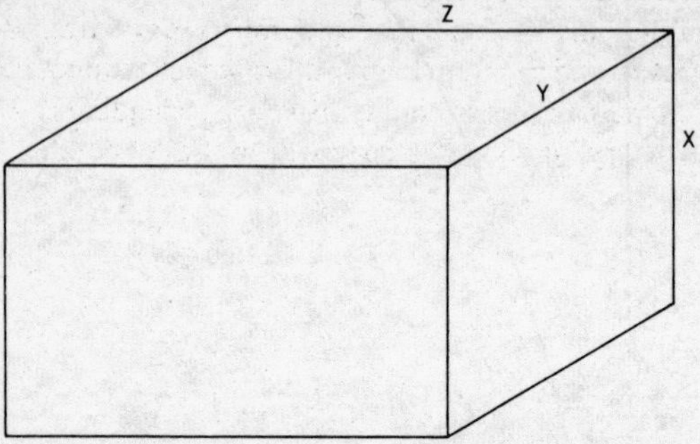

Parcel post: the largest volume

$$V(\text{volume}) = xyz; v = x^2(10-4x) = 10x^2-4x^3$$

at one end, keeping the length fixed. To get the maximum volume we shall need, whatever the length, to have the area of the end as large as possible. Again, it is useful to appeal to one's intuitions.

There seems no good reason why either x or y should be bigger than the other; intuition tells us a square would be best. We will settle for intuition remarking that we could prove it using the method employed with the allotment. So we no longer have x and y. Both of them are x, the girth of the box is therefore 4x. This allows us $10-4x$ for the length if the total is to be 10 ft. Now our expression for the volume V only has xs in it, and we are in business. All we need is to differentiate it and put it equal to zero.

$$V = 10x^2-4x^3$$
$$DV = 20x-12x^2$$
$$x(20-12x) = 0$$
$$\text{so } x = 0 \text{ or } x = 20/12 \text{ or } 5/3.$$

In a few lines of working we know that the box is $1\frac{2}{3} \times 1\frac{2}{3} \times 3\frac{1}{3}$ and that this is the largest possible. A touch of magic about that.

The examples of maximum values that we have given are comparatively simple, though the second answer would be less easy by other means. The determination of maxima and minima can be of direct practical use in industry and commerce. In our chapter on 'Mathematics in Action' we describe a two-stage manufacturing process with storage necessary between them. When we set up a model of this we get equations connecting the time and the cost. These equations contain things we do not know (variables, like the sides of the box) but by differentiating we can determine maximum or minimum conditions. The uses of calculus are extremely varied and important.

We see calculus then as the study of continuous change. When one quantity varies with another, speed with time, volume with length, it is the calculus that we turn to. It arises first in the study of motion, applies to a wide variety of graphs, links the fundamental numbers in an extraordinary relation, copes with growth and decay, maxima and minima. It cannot deal with abrupt changes, what we call 'discontinuity'. In the Appendix (p. 237) we discuss 'catastrophe theory', designed to cope with just that situation.

9

Taking a chance: living with uncertainty

You have been tossing a coin in a series of matches to decide who starts. So far you have called 'heads' five times in a row and lost every time; what now is your policy when you toss for the next time? People react in various ways, and are ready with reasons for their decisions. The most common is to continue to call heads on the grounds that it cannot keep coming down tails, and that by something called 'the law of averages' heads and tails must balance out eventually, so it is time we had a head. This is a totally unreasonable position, yet the fact that so many hold it illustrates the difficulty we are going to have with probability.

There are two tenable positions in this situation. There is the one based on very slim evidence, but evidence none the less, that the coin has a bias, and tends to come down tails rather than heads (even if not deliberately constructed that way). If you believe that this might be so, your immediate policy is to call tails.

If, however, different coins have been used, or you are confident that the coin is standard, then there is the simplest of all policies; you have no policy, and call what you like.

Within this simple situation lie some of the crucial points in probability theory; let us tackle them one at a time. The belief that if you toss an ordinary coin a great number of times there is some 'law' which says it will not be far off equal in the balance of heads and tails, is justified. It is, for instance, extremely unlikely (but not impossible) that in tossing a coin a hundred times there would be a hundred tails or heads. (Against this is the other fact that if you go on tossing for as long as you like, there will be a hundred of one kind in a row at some time.) Given a large number of throws we are very likely to get somewhere nearer 50/50.

Why then is the first position so unreasonable? The unreason lies in the assumption that somehow the coin knows it has come down tails five in a row and is determined to balance it out. Or even in the belief that an outside agency (some god of chance?) is ensuring that balance will be reached. Once a situation is passed, that is it – it is passed. The fact that five tails have just occurred cannot influence what happens next. If you believe otherwise you believe in magic; so do not read on.

There are situations where each choice or happening affects the next. If from sixteen people we pick a team of eleven, one at a time, then each successive choice alters the group from which we are choosing. So the different 'events' (the choices) are dependent on previous ones. That is not the situation with the coin. The issue as to whether each event depends on previous ones or not is extremely important. We must be clear which situation we are in.

The second policy, picking tails on the grounds that there may be reasons for tails turning up more often, also leads to a central issue. It is easy to assume that the words 'equally likely' in relation to a coin coming down heads or tails are easy to understand, yet there are two ways of looking at it. The first way assumes a physical law that makes a symmetrical coin fall equally one way or the other. The second does not assume such a law, but says that the balance of heads and tails rests on the experimental evidence of a great many previous tosses. The (thin) evidence here is that the coin comes down tails more often.

If we have no experimental evidence, or have different coins for each toss, then the mathematicians' logic cannot be escaped. Bet which way you like, and accept that it is guesswork.

If you feel already that your intuitions about probability may not be accurate, you will be in good company. In 1754, a very great mathematician named D'Alembert was presented with this problem:

'In two tosses of a single coin, what is the probability that heads will appear at least once?' He reasoned that there were three cases, two heads, two tails and one of each and that the probability was $\frac{2}{3}$ or two chances in every three. An easy trap to fall into; the cases, however, are not 'equally likely'. Below we show the four equally likely outcomes of the two tosses:

H H T T
H T H T

and of the four cases, heads appear in three. The probability is therefore $\frac{3}{4}$ not $\frac{2}{3}$. D'Alembert would have been well advised not to enter gambling in a big way.

The theory of probability has its roots in gambling a century earlier when the Chevalier de Méré, a well-known gambler with an interest in mathematics, posed the problem to Blaise Pascal of how the bets should be shared if a game was left unfinished. This led to a spirited correspondence between Pascal and Fermat which was the first development of what has become one of the most extensive branches of mathematics.

Let us bring this up to date with an unfinished five set match at tennis. We have to make the rather unlikely assumption that the two players are equally matched, or that each is equally likely to win a set. Altogether rather too many assumptions, but from there let us suppose the match is left unfinished one evening with one person leading 2–1 at some stage. What odds would you offer on the player who is leading? This corresponds with the way you would split the prize if the match were left unfinished, which was the Chevalier's question.

The next set is equally likely to go to A or B; the probability is $\frac{1}{2}$ each:

A $\frac{1}{2}$ and he wins
B $\frac{1}{2}$ and the score is 2–2.

Thus the chance of A winning is already $\frac{1}{2}$. If we now play the last set, assuming B wins the fourth, then this $\frac{1}{2}$ chance has to be equally split:

A $\frac{1}{4}$ wins
B $\frac{1}{4}$ wins.

The whole situation now shows A to have a chance $\frac{1}{2}+\frac{1}{4}$ or $\frac{3}{4}$ and B has $\frac{1}{4}$.

The odds are 3–1 on A, so now place your bets. There is an interesting exercise we could do to introduce an experimental element into this rather formal problem. Buy a book recording the results of all the major world tennis tournaments, and pick out all the five set matches where the score was 2–1 at some stage. Then check the proportion of times that the person leading at that stage did in fact win. In doing so, we ignore their ranking and likely chance of winning against that particular opponent, and simply rely on statistical evidence.

We have used two ways of writing probabilities. We said that A's chance was $\frac{3}{4}$ or that the odds were 3–1 on him. The statements are the same. The first says that A wins 3 out of every 4 and the second that he wins 3 and loses 1.

Games and gambling often rely on the fact that the odds are known. Roulette and vingt-et-un are games of very little subtlety. With a single zero on the wheel you are paid off at 35–1 in a situation where the odds are actually 36–1; so gradually you lose. 'Winning' systems eventually always founder on this basic fact. Only if the wheel is not balanced can a system detect it and capitalize on it. With vingt-et-un, commonly called 'pontoon', the simple fact is that the banker wins if your hands are equal and so wins altogether in the long run. We are back with our law of averages, on which the casino owner and the bookmaker rely. They can be beaten, by chance. There is no rule in the law of averages that says black 7 cannot come up five times in a row with you backing it. If in any real danger, the bank claims it is 'broken'. It has rules for how much it can lose, but not how much it can win.

Other games are more subtle in their odds. The basic game of poker involves a hand of five cards, some hands being stronger than others. The players bet on their hands until all have bet the same amount (or dropped out) and the strongest hand scoops the pool. Much of the skill depends on the ability to bluff one's opponents and force them out even if they have stronger hands, but underlying the subtleties of human interaction, there are hard facts as to the likelihood of holding certain hands. It is calculable how often you may hold all cards of the same suit, or have three cards all of the same number, and the ranking of the hands reflects this underlying chance of holding them.

Bridge too bases its bidding on likelihoods of gaining a certain number of tricks out of thirteen. If one has slightly more high cards than average, and rather more cards in some suits than average, the probability is that you may make more tricks than the opponents. In play there are some simple examples of probability. If you lead a small card to a holding of Ace, Queen, with no idea where the King is (except that opponents have it) then playing the Q gives you a probability of $\frac{1}{2}$ of making that trick. Many of the odds are far more subtle.

Games like craps depend on throwing dice. If you throw two symmetrical dice the sum of the faces may be anything from 2 to 12 but, as with tossing a coin twice, not all numbers are equally likely. Look at our table:

6	7	8	9	10	11	12
5	6	7	8	9	10	11
4	5	6	7	8	9	10
3	4	5	6	7	8	9
2	3	4	5	6	7	8
1	2	3	4	5	6	7
	1	2	3	4	5	6

2: 1	8: 5
3: 2	9: 4
4: 3	10: 3
5: 4	11: 2
6: 5	12: 1
7: 6	

Once again, pattern is evident both in the table and in the listing at the side of how many times each number occurs. It shows that there are not eleven equally likely outcomes (2 to 12) but altogether 36 outcomes, all equally likely, but such that a particular total may be much more likely than another. 'Snake's eyes' or 2 only comes once in 36 throws, as does the double six. A total of 7 is six times as probable, there being six outcomes in thirty-six which give it.

An interesting exercise is to examine the pattern of possible numbers when the numbers on the faces of the two dice are multiplied, not added.

Before we move away from the realm of games and gambling let us look at four problems, on which you can test your intuitions about probability. We shall first say what they are, then give the answers. They will almost certainly prove surprising, and in the Appendix (p. 238) we shall show the working that arrives at the results.

(1) There are thirty people at a party. What odds would you offer against two of them having the same birthday?

(2) Place two packs of cards face down and turn them over, one from each pack at a time. What is the chance that you will get a perfect match in one run through the pack (i.e. number and suit)?

(3) The odds against 'snake's eyes' at dice are 35–1. Suppose you throw the dice 36 times. What are the odds against you not getting 1 : 1 on one of those throws?

(4) A gambler offers you three dice, with the normal six sides. You observe that they do not have the numbers 1–6 on every die, but have a single number on every face. The same number may appear on more than one face. He now offers you the following game: 'Choose a die and throw against me at £1 a throw. The higher number wins but equal numbers is no bet. At any stage if you do not like your die you may take the spare one or mine; I can then choose and we roll again.' Should you take this offer?

The first answer is that it is 7–3 on that two will have the same birthday. When there are as few as 23 the odds are still even (probability $\frac{1}{2}$) that two will share a birthday.

The second again is odds on. Many feel that it is long odds against, but simply try it.

There is a common belief that if something is say 35–1 against, you can guarantee a favourable result if you try more than 35 times. It is not so. There is a 36% chance that you do not throw 'snake's eyes' in 36 throws.

You should not take this offer. The gambler knows his stuff. If we call the dice A, B and C, you will find that A beats B (in the long run), that B beats C and C beats A! His simple rule is that whenever you choose a new die, he takes the one that beats it.

These results do not accord with most people's intuitions. Yet they can be backed up mathematically, and for the many people who do not believe things simply because a mathematician proves them, they can be confirmed by experiment.

It is no bad thing to know what you are about when gambling, but there are issues of greater importance, and probability theory has come to play a larger and larger part in our affairs.

Probability has proved a study that appears to impose some order on apparent randomness, and to enter into a range of human activities. The first apparently random result is what month your birthday is in. Here our intuitions are likely to be sound. Since our birthday can be any month we are inclined to believe they are all equally likely and that a birthday graph of a large population might look like our diagram opposite.

That is how it should look if every month were exactly as likely, yet somehow we do not expect it to come out exactly flat, as we have drawn it. We expect flatness but not perfect flatness. Then again, we might find that the birthday graph, when we get the actual statistics, showed a decided rise in certain months, perhaps owing to climatic conditions nine months earlier.

Our heights too are random, but there is order in the randomness, though this time the pattern is different. If we take the adult male population we do not expect there to be the same number of men of 5 ft 2 ins as of 5 ft 8 ins,

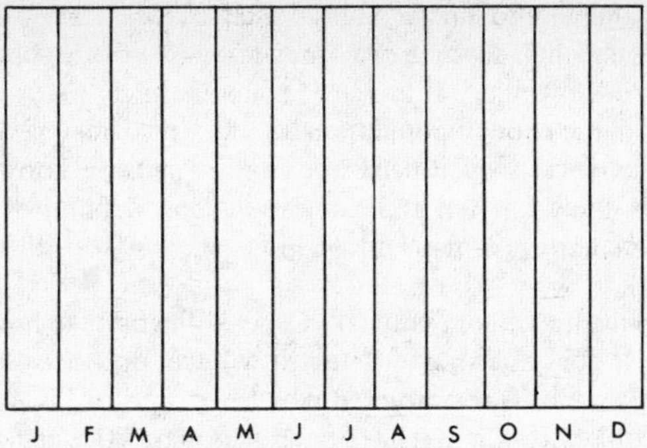

A birthday graph

though it might be the same as those of 6 ft 2 ins. We expect, from our experience, there to be many more men near average height (say 5 ft 8 ins) than well out to one extreme or another.

Originally we would build this graph, too, in columns, but as we put the height ranges closer and closer, and the number of people greater and greater, we begin to approximate to the bell-shaped curve we have drawn. This is perhaps the most important curve in probability, for many apparently random matters fall into this pattern.

This gives one pause for thought. It is no surprise that in formally constructed systems like mathematics there should be many patterns, but why the world should exhibit pattern of this sort is a vexing question.

Naturally such distributions are important for us to know commercially. It is unlikely that the rag-trade makes a study of probability, but the clothes they supply need to fit a population most of whose dimensions fall into the pattern of the 'normal distribution' illustrated in our graph of heights.

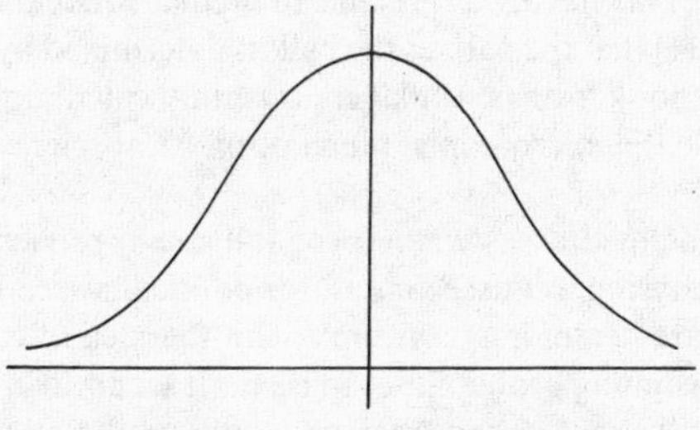

Height graph: the 'normal' curve

Chance enters into many issues – such as when we die. We cannot accurately assess when any one of us may leave this world, but we have very good notions of how many of any age group are likely to die in a particular year. Given a large enough population and leaving aside major disasters, or the end of the world (which need not concern us), we know how the age distribution is going to alter, and can give in some detail the life expectancy within particular groups of people. This is the basis of all insurance and actuarial work.

When an insurance company takes a bet with you as to how long you are going to live, it does not know whether it will win or lose (it loses if you die early – you win). Nor is it concerned whether it wins or loses with you. The number of people with whom it has similar bets makes the insurance company sure of its eventual financial gain.

It may sound macabre to talk of betting on one's life, yet that it what life insurance is. There are many other types of insurance; many of us have car insurance as well as life insurance. When we go abroad we insure against many things. The basis of everything is betting against what happens in the future, and always the organization arranging the bets, like the bookmaker, is going to win out overall. Yet there is, of course, sound sense in insurance. It is likely that the money you pay to the car insurers is a great deal more over the years than the amount they have to pay out for your crashes. Yet you cannot afford not to have it, for a bill could arise that you cannot meet, and will produce total financial ruin. It is this against which you insure. Effectively that big risk (say £100,000) is infinite to you, so that straight mathematical calculations are inappropriate.

Calculations might go like this. Suppose you pay over 20 years against a single large risk of this sort. That would make £5000 a year to balance the risk (ignoring inflation). If then the chance is only one in a hundred overall, this would mean £50 per year. Suppose the company asked £100 per year of us. We might feel it unreasonable, but the fact is that we cannot take the risk, for a number is effectively infinite when beyond our capability.

Probability is closely intertwined with cause and effect. In simple situations, like balls colliding on a billiard table, cause is followed by a clearly defined effect. We may not always be able to predict where they will finish after a stroke, but we feel that the path of the balls is determined by how the first one is struck. Some snooker or pool players certainly calculate with remarkable accuracy. Outside the microscopic, much of physical science seems to be like this.

It is in less defined areas where probabilistic approaches become more important. In analysing trends, economic or political, we cannot say what will happen, but we are beginning to grapple with the idea of considering all the outcomes, and assigning probabilities to them. It seems likely that probabilistic approaches will develop and offer an alternative to statements assigning inevitability to certain economic events, as is characteristic not only of

Marxism but other political beliefs. Karl Popper gathers together various social philosophies which raise claims to firm predictive ability and refers to them as historicism. His main objection is against the assertion by some that we are at the mercy of these trends, and cannot resist, but he also makes a clear distinction between scientific prediction and historical prophecy. We remain very far from accurate predictions, derived from history, of our future; but we can begin to draw some suggestions of what may lead to what, and even to calculate some likelihoods. But the main purpose in doing so might surely be to avoid certain outcomes at all costs.

One of the most frightening of the activities of the futurologists is the discussion of likely situations 'after the bomb'. Detailed probabilities of survivors and facilities left have not proved encouraging; it can be even less encouraging to think that people are looking to such a time.

These are serious matters; let us leave them and ask another question: 'Where do you sit on a bus?' Try this as an exercise with any group of half a dozen people and you will find there are many preferences offered. There is the question of upstairs or downstairs, the left or the right, whether the seat is part occupied, what the window is like.

Our own personal computers, carried in our heads, assess all our preferences as we enter the vehicle. Observe yourself next time you do so. Certain of these preferences may be simply overwhelming, and dominate the others, but usually we are faced with a series of favourable and unfavourable choices that lead us to sit in a certain place.

All decision making is like this; the bus seems an amusement, but typifies our way of working in small or large issues. We attach various weights to certain features of a situation and these create probabilities as to what we do. It is seldom easy to analyse. Imagine yourself faced with the plan of the top deck of a bus and trying to guess the order in which seats fill as the passengers get on one by one. We believe perhaps that the first person entering the top deck is likely to sit at the front left or in the seat with an armrest at the back left. Both have high probabilities, but every single seat has a probability, perhaps less. Assessing human behaviour, in crowd control or in marketing sweets, is a tricky task but it is increasingly being attempted. We have no certainties, only a variety of likelihoods.

In life we are often amazed by coincidence, and are too often inclined to offer magical rather than rational solutions. If we find we meet someone after a long time, and shortly before had been thinking of them, we are inclined to credit ourselves with some form of prescience. Suppose an event has indeed a low probability, say one in a thousand. We are normally greatly surprised when it happens, and this is realistic. However, once we consider how many events have occurred to us that day, the surprise lessens. If we have seen a thousand events of that particular type, even over a considerable time, we will still be surprised that it happened. Our calculations on the dice are matched here; it is still likely at the 60%+ level that we shall see it happen.

The chance of getting a nine card suit at bridge is very low (how low is left as a problem) but if you play bridge often enough for enough years, you will get one. It would be very surprising if you did not. Isaac Asimov puts it succinctly in this way: 'It would be very unexpected for the unexpected not to happen sometimes.' This goes back to our experimental approach to probability. If something happens once in a hundred times, you expect to establish that fact by doing many hundreds of trials and see it happen that number of hundreds. We are still inclined to think that low probability events should not occur; and that is a nonsense.

We started by tossing a coin, became involved in world futures (or non-futures) and the vagaries of human beings. Probability penetrates many areas; does it have personal implications for how we behave? Certainly anyone who takes a probabilistic view of life can be better fitted for the range of decisions big and small that we all face every day. In deciding what we do at any of the branching paths that constantly open up before us, we try to assess the probability of a favourable outcome if we take one path rather than another. To do this we may need to delay decision until we have all the information we can have before this decision needs to be made. So we wait to get this and do not rush into a decision, but we make it when we have to. The trauma of wanting to be sure that what we are doing is right is not one that the probabilist gets involved with. He or she knows that certainty is not possible. Nor does it matter if our information leads us to believe that one course is only favourable on a 60–40 balance. You take the 60% route and do not fuss. That is the right and sensible thing to do.

Hindsight often shows our decisions to have been 'wrong'. It may be that the chance was 60–40 and the 40% came up – as it must do 40% of the time. The decision was not wrong, even if the outcome was not favourable; the things are not the same.

Hindsight might reveal that the odds were in fact 40–60, not as we thought. If that could only have been known at a time after the decision was made, we were still not wrong. Odds depend on what you can know.

The message is only to have regrets if you have failed to take into account things that you could have known before the decision, or that you wrongly judged what you did know. It takes a good deal of internal security to live with uncertainty, but it can be quite comforting when achieved.

10

Modern mathematics

The term 'modern' mathematics gained currency some years ago, largely because many adults became aware that their children were studying aspects of mathematics, sometimes even in primary school, of which they themselves knew nothing. This occasioned disquiet, not only because of its unfamiliarity, but because the purposes behind it were not clear. 'Modern' often embraced both content and method of teaching; here we shall discuss only content, and only two aspects of that.

Mathematics is often thought of as a fixed and complete body of knowledge; nothing could be further from the truth. In our next chapter, on 'Mathematics in action', we shall see something of two recent developments, operational research and graph theory, but the two best-known topics arising from school mathematics are 'sets' and 'matrices'. To call them modern is stretching a point, since both have been within mathematics upwards of a century, but since much of the mathematics we know was familiar to the Greeks, we might at least accept the term as comparative.

Mathematicians are apt to start with definitions or propositions and move forward from there. This can be singularly unhelpful for the layman, and could account for some of the distaste felt by some for the subject. We could start with a definition of a set as 'any undefined collection of objects'. If then we start to postulate relationships of a formal nature between these objects, and then manners in which they operate on one another, we seem to be in a world somewhere between Lewis Carroll's and that of Kafka. Bertrand Russell neatly summed up this approach: 'Thus mathematics may be defined as the subject where we do not know what we are talking about, neither do we know if what we are saying is true.'

It is easier in talking of sets to start in our own experience. All of us have had a great deal of experience of sorting and classifying. Certain types of collecting, such as stamps, can lead us into quite sophisticated categories, and many everyday activities lead us to think carefully about where something may be found. Shops contain collections of objects, and so meet our definition of a set. The shopkeeper usually has clear ideas as to what belongs in that set and what does not. A fish and chip shop does not normally sell wine, even if its customers might like to drink it with their meal.

Suppose we wish to buy a safety-pin; consider what sort of shop might sell them. If there is a haberdashers at hand, we would feel confident in asking there, but if not, should we search the shelves of a supermarket, ask at the chemist's, or find a hardware store? We are busy forming sets in our minds, looking for collections in which we might find a safety-pin. This is easy enough to understand, but why should we dignify this commonplace activity with a name from a branch of mathematics? We hope gradually to make this clear.

Many occupations involve sorting, and an ability to do it efficiently can be very important. The small shopkeeper not only decides what he will stock, but must have a good system for finding items. Customers do not enjoy waiting while the retailer searches hopelessly in a muddled storeroom. It can particularly matter if the goods are perishable. At all levels in the retail trade, sorting is relevant; the girl on a counter in a department store, the warehouseman, the clerk at central office keeping accounts, all need their different levels of understanding of the ways their products are sorted and organized. Similar skills are needed in running a ticket office, be it for the railway or the theatre. The efficient car mechanic sorts and orders his tools, reaching for them as he needs them in a place where he knows they will be, as does the sister in a surgical theatre.

It is the common perception that the value of mathematics in the everyday world lies in calculation. It is useful, but so is sorting. Yet we feel that sorting is a matter of common sense, and that is something we either have or do not, God-given rather than learnt. The tendency is to learn on the job, either accepting the way it is already done, or for the self-employed making a system and then sticking with it. The trouble then comes when anyone tries to improve that system, even if it be to the great advantage of the enterprise.

Let us move to more complex, yet still everyday activities. The natural one to start with is libraries. The activity of sorting books is one which young children enjoy and which raises interesting problems for them. They classify in many ways:

'Books I like reading'
'Red books'
'Paperbacks'
'Big books'
'Adventure stories'
'Bedtime books'
'Information books'

and so on. The classifications come naturally to them and they make sense. Yet when they try to put the books in piles using these labels, obvious difficulties arise. Into which pile should the big red paperback go? With a small number of books, or even a home library of several hundred, you can get by without too much organization. The mind can cope simply by remembering, and even if we cannot find the book we want, it may be fun looking. Beyond this level, however, mere common sense and one's personal memory are not enough. Librarianship, and within it the question of information retrieval, involves complex sorting processes and sophisticated analysis. Where exactly might we find a treatise on church rituals in Persia in the twelfth century (if indeed there were any)? Would we start in a religious, geographical or historical mode?

As we become more and more computerized, the retrieval of information becomes easier for us, but not always for those who feed it into the memory banks. If the police follow and stop you in your car, you may find that when they do so they are able to address you by name. All car numbers are stored in computers, and the name and address of the owner is readily available. Maybe they are listed alphabetically, but is there also some area-based sorting? Certainly in keeping criminal records it would not be very sensible to list everyone alphabetically, with crimes from parking on a yellow line to multiple murder. There must be a lot of classifying, and many cross-references. The quality of such record-keeping may determine whether or not a Yorkshire Ripper is caught.

The point about all these common sorting activities is that they are all similar. (Maybe they form a set!) Perhaps the man running a large motor accessory shop does not see himself as a librarian, yet he shares some skills. A person who looks at sorting activities themselves may not be concerned with whether what is sorted is rare books or crankshafts. We are back at our first starting point, then seen as unhelpful, that a set is any collection of objects. At times it may matter what you are sorting, but at others it may merely be distracting. The process rather than the material may be the significant issue.

Properly to understand the relevance of sets – collections of objects organized in different ways – we need to approach the idea in a variety of ways. Our first was through everyday activities, our next will be through one feature of the scientific method. Chemistry may be said to have started with our recognition of the difference between metal and stone. With our natural belligerence, we discovered that metal was even better than stone for hitting other people. Pause for a moment to think of metal and stone. We all feel we can distinguish them, and given a mixture of chunks of each could sort them out. It is true that we could, provided the material was sharply enough defined, but could we say what made one metal and one stone? Perhaps not; that is where chemistry starts.

The study of the subject received its next great thrust from Man's cupidity and his desire for personal survival. He sought the 'philosopher's stone' and the 'elixir of life'. The first would transmute 'base' metals into 'noble' gold (note the social analogies) and the second would bestow the rather doubtful blessing of eternal life. (As chemistry began to develop from alchemy these hopes were seen as foolish and treated with scorn. There are often twists to these tales. The changing of one element to another has now been established.

Transmutation, though not of lead to gold, is an accepted fact. We also have medicines which prolong life, even if they do not give immortality.)

Chemists began to distinguish elements from compounds (an important distinction), and then to observe among certain elements some common properties. We have said that, mathematically, a set can consist of any collection of objects, yet sorting by attribute seems to lead to the most purposive advances in understanding. The chemists began sorting the elements into sets. As more and more elements became known, they fell into families. One such family is now a household word – the halogens. They include fluorine, used in our water and, it is claimed, by toothpaste manufacturers; chlorine, compounds of which are used in our swimming pools but whose first main use (typically) was as a weapon of war, when this heavy dark green gas rolled across the battle-fields during the First World War; and iodine, which in various forms played an important role in the cleansing of wounds before the advent of modern antibiotics.

These substances are seen to have family resemblances, perhaps rather deeper than the contrast between metal and stone. This grouping process continued until it led to the 'periodic table' which placed all the elements in order in such a manner that families like the halogens could be seen to belong together. A part of this table, containing the halogens, can be found in the Appendix, p. 239. It was this table, arising from a sorting process, which led to speculations about the internal structure of materials, and the eventual capacity to make bombs which may yet destroy us all. Sorting is not the only process at work, yet it is an essential one; the end of such a line of development is not easy to forecast from its beginnings.

Another scientific area offers further evidence of the power of sorting. Darwin started in a way that seemed to differ little from the collection of leaves from various trees, or the activities of the dedicated bird-watcher. He simply tried to sort what he saw. Though in mathematics we may regard it as of little consequence how we sort, in science it is clear that some distinctions are more important than others. We might at random pick number of legs as a criterion for sorting living things. It happens to be important. Less obvious is whether an animal is vertebrate or not. Why should that prove significant? Yet it is. We could sort out living things which fly, and which ones live in the sea, but that seems less helpful, for we have birds which do not fly, and we prefer not to classify the whale with other sea creatures, but with mammals, who suckle their young. We seem to be in the position of the children classifying books. Some looked at the concrete object, the book, others at its content and meaning. Deciding what is important in classifying is not easy, but Darwin clearly had a knack of knowing what mattered. He sorted the animal kingdom; once it was done we wondered why no one had done it before. Once the animals had been sorted and ordered, it became apparent that there were certain families, as with the elements. Nowadays our relationship to the great apes is obvious to all. We really only needed to look. It need not imply that we have a common ancestor, but the fact that we belong with certain animals rather than others must be evident. Yet at that time it was firmly denied, with that vehemence the truth always provokes, and with a degree of personal animosity to those who shared Darwin's view that shakes one's optimism about human tolerance.

One of the most important bastions that evolutionary theory attacked was the assertion of many religious people that we were not even to be classified with the animals. This making of sets was not to be permitted; it was an attack upon God. Most, not all, churches have adjusted to accommodate Darwin's ideas, but evolution as a general principle, let alone issues within it still open to doubt, is still rejected by those who believe the world was created in 4004 BC.

The point of the whole story is not that conflicts may arise between the various philosophies that men hold, but that a decision to sort and order the animal world should lead to a major confrontation such as happened in the last century. Darwin, looking at flora and fauna in a scholarly and detached manner, does not seem a likely starting point for an attack upon the major religions. The moral is that sorting is a potent operation, and where it may lead we cannot tell.

Our next approach to sets takes us to one of the central concerns of philosophy – objective knowledge. Elsewhere we shall deal with the role of mathematics in questions of uncertainty; here we note that in seeking some philosophical starting point it was natural to look at the counting numbers. Kronecker, a great mathematician, once said: 'God made the integers; the rest is the work of Man.' It therefore became, about the turn of this century, a major task in philosophy to establish the whole numbers on a firm basis. There were strong reasons for supposing that if this could be done then the rest of mathematics would be secure. Two approaches were made to this task. Peano developed the natural numbers from the idea of a successor. Once a starting point was established, then the rest of the numbers would follow. This leads to ordinal numbers, first, second, third. . . . The approach that interests us here is the one put forward by Russell and Whitehead. They started with undefined sets of objects, and used the idea of pairing off; this establishes a one-to-one correspondence between the objects of the two sets. It seems a simple start. In the first of our diagrams on p. 144 a complete one-to-one correspondence is set up; in the second it is not. This compares without counting, and takes us back to a pre-number stage, which is not a trivial step. We use it often; we observe that a theatre is less than full, exactly full, or over-full, with people standing. This does not involve counting. The cricket umpire moves one stone every ball bowled; it saves counting.

In our next diagram we find a correspondence set up between several sets. Whatever the elements of the sets may be, the thing that they have in common is characterized by this correspondence, and we call it the cardinal number of the set. We are now able to name this number and call it 'five' if we wish.

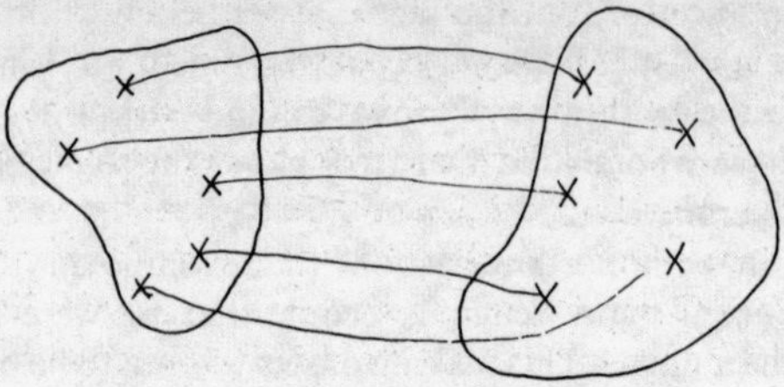

One-to-one correspondence (1)

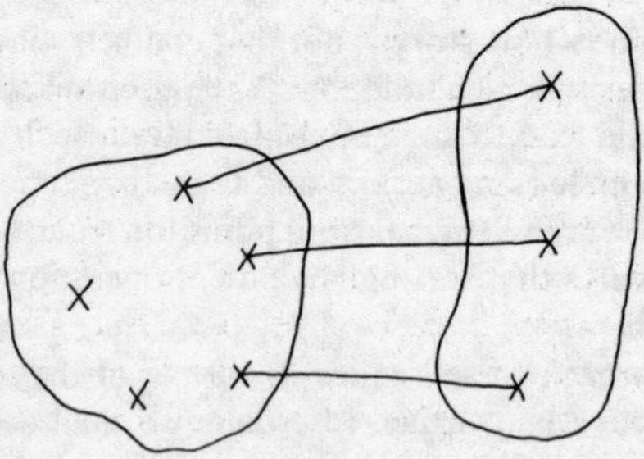

(2)

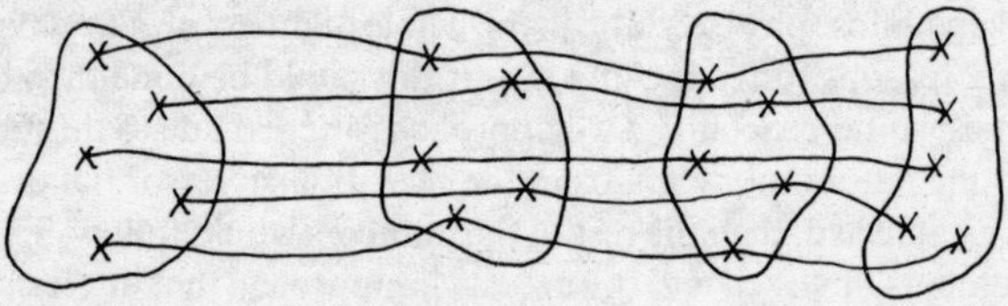

(3)

Mathematics, tackled like this, sometimes resembles a medieval disputation on the number of angels that may stand on the head of a pin. Our early familiarity with the counting numbers prevents us seeing any difficulties within them, yet there are such difficulties, and they have exercised some of the most powerful minds of all time. It is now common practice in infants' schools to approach number in this way. Children are encouraged to collect sets of five objects (paint-brushes, pencils, chairs, friends, what you will) to establish the 'fiveness of five'. The fundamental concerns of the mathematical philosophers are mirrored in the infant classroom. It is at this stage that the bases are laid. We need constantly to consider how it should be done.

For our next point we stay with young children, and look at the beginnings of language. It is language that most marks us off from the rest of the animals. Experiments with apes have had some remarkable results in establishing a limited vocabulary, but it does not compare with the language acquisition of the pre-school child. There are various views as to how a child gains so much language so quickly. One of the most important is that put forward by Noam Chomsky, whose studies in linguistics led him to see an underlying structure common to all languages. From this he was led to the belief that there was some form of genetic imprinting, some patterns in the brain which allowed language to be learned so quickly. Even more recently, with the growth of computers, comparison has been made with the circuitry which permits a machine to accept its own particular language; such comparisons should not be pushed too far.

Whether you accept the Chomsky position or not, an important function of our brains is receiving input, sorting it and then giving it a name. That is where language begins. Consider the way a child acquires the word 'cat'. We are not concerned whether he or she can read or write it; that comes much later. Can the child identify a cat, understand the spoken word for it, and then perhaps attempt the word? It will have been necessary to see a number of cats, and, equally important, a number of not-cats. In mathematical terms we are establishing a set and its complement, as we indicate in the diagram.

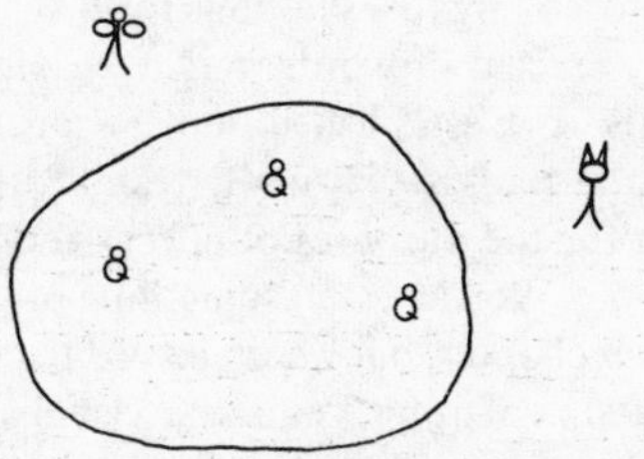

The cat

As an adult, despite our belief that we know what is a cat and what is not, we might be pushed to give definitions rather than examples to an intelligent being who had never seen one. Perhaps a totally alien being might classify quite differently anyway. As for ourselves, we are later in life obliged to widen our earlier category of cat to admit lions and tigers.

The process of making sense of our environment relies upon sorting and ordering what we see, hear, smell, taste and touch. An essential part of this process for the human being is the language that goes with this sorting and ordering. Its function is to stabilize these processes and make them capable of being communicated to others. There seems to be an intimate relationship between early concept development, sorting (a mathematical issue) and naming (the start of language). There is ground for supposing that our psychological processes, early mathematics, and language are closely knit. This has important implications in regard to the nature of mathematics, which we shall discuss in our last chapter.

Our next point about sets relates to 'intelligence'. What it is and how it develops is a topic fraught with rational difficulties and bedevilled by emotional responses, and we shall not discuss it here except in the context of sets. The ability to sort is clearly different in different individuals, whether for genetic or environmental reasons. The ability to perceive samenesses and differences seems to be one aspect of cognitive power; put it no higher than that. This amounts to an ability to sort – to know what goes with what. Given any material to sort, some will see more categories than others, make more distinctions. There is a problem with those who do not see a particular distinction, for if they cannot see it they may believe it is not there; it is a 'catch 22' situation. Within science the increasing understanding of sameness and difference is a potent force for advance. Some experiments with children are described in the Appendix (p. 240).

Our next discussion about sets remains in language and has a distinctly English flavour. We shall look at Boole and Venn, and illustrate with examples from Lewis Carroll.

George Boole was born in 1815, and in 1854 he published 'The Laws of Thought'. So unusual was this that it was not even thought of as mathematics. He sought to reduce reasoning to a symbolic form. This meant that processes of thought, normally carried in words, might become the manipulation of algebraic symbols. His work established Boolean algebra and opened up vast new tracts for mathematicians to play in. Bertrand Russell (occasionally given to mild exaggeration for the purposes of effect) said that pure mathematics was *discovered* by Boole. We have seen something of the apparent oddities of his algebra in an earlier chapter, but we shall need to develop this further, and we shall use the diagrams of John Venn as a visual aid in our use of sets to interpret certain language statements. In the diagram at the top of p. 147 we start with some numbers, not because they are special to this topic, but because they provide a simple example.

In the enclosure on the left (its shape is not important) we have a collection of numbers divisible by 2; on the right an enclosure with numbers divisible by 3. It is not difficult to work out that the numbers in the overlap are divisible by 6, since they are divisible by 2 and 3.

We look now with special emphasis on one word – 'and'. The numbers divide by 2 *and* 3. We seem to be making much of little (again?). Yet attention to such words can be critical. Let us go to law for a couple of examples.

The definition of stealing involves the words '... and with the intent permanently to deprive the owner ...'. It may not seem natural to pay special attention to small connecting words such as 'and', but that is where Boole has taught us to look. Suppose you walk out of a shop carrying something you

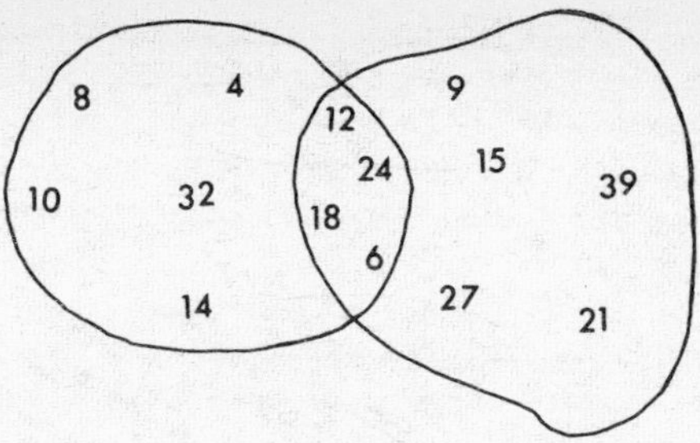

Dividing by 6

have not paid for. In our diagram below we enclose such actions in our oval labelled W; this means you walked out. Many who are then challenged exhibit signs of guilt and confusion; they may be the very people who are not thieves. Regular shoplifters are probably more organized in their responses. In the right hand oval we put those who had the intent permanently to deprive the shop of some of its goods.

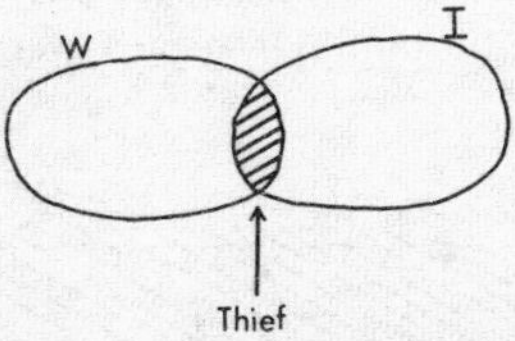

Shoplifting

Now we come to the point of the 'and'. Only if you are in the overlap are you guilty. You must have taken it 'and' intended to keep it (or sell it). The honest but absent-minded are in the left hand part, the criminally minded but incompetent in the right hand side, and the thieves in the middle. If you are one of that left hand lot, beware – you still have to convince the magistrates.

Legal and quasi-legal statements are a minefield of logic. London Transport offers us this statement: 'It is an offence to travel on the Underground without a valid ticket and with the intent to avoid payment.' The same diagrams, but different labels. Those on the left in the top diagram on p. 148 do not have a valid ticket (NV), those on the right intend to defraud (I). Again you need to be in the overlap to be guilty, but here there is an added subtlety. Is it possible that, intending to defraud, we have a valid ticket? Surely not. So our diagram should look like our revised model, with I inside NV. We now see that the outer ring is pointless. You are guilty if you intend to defraud, and that is that. The first phrase is now seen as a ploy, to persuade you that you must have a ticket, even when you cannot get one at their booking office.

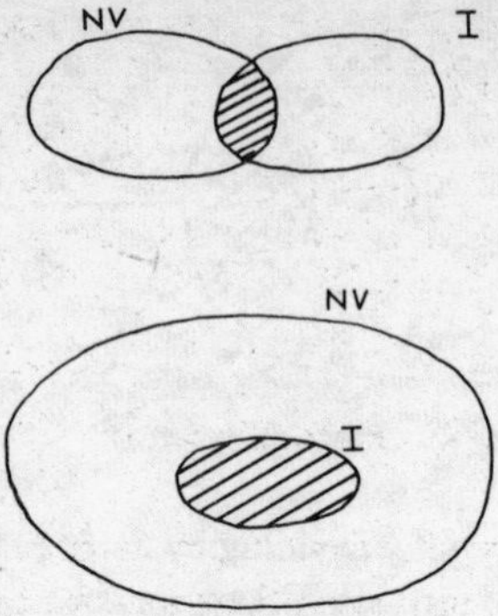

Underground travel

The next logical connective is 'or'. At one time British Rail offered railcards (permitting cheap fares) if you were a student or a senior citizen. We have our overlapping ovals again, on the grounds that you might be both. The shaded area covers all those entitled to a card. Since it covers the whole region the overlap is not now important. 'Or' means everyone in the two sets.

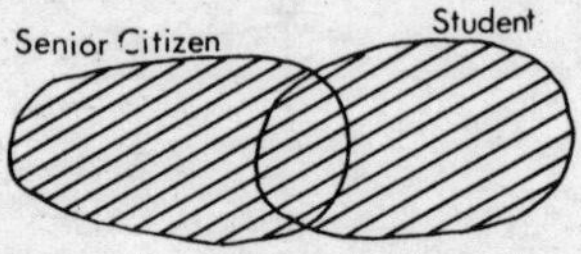

Railcards

Now for 'only' and 'all'. Consider the two statements:

'All fools play football.'
'Only fools play football.'

We are not asserting the truth of either, merely comparing what they say. They illustrate one of the commonest invalid conclusions drawn by people. The first statement does not imply that if you play football you are a fool; the second one does. It is nevertheless unwise to make the first statement to a keen footballer lest you do not have time to explain the logic. The two diagrams opposite make the point. In the first there is a shaded area where we may put footballers who are not fools. In the second all the footballers are fools, and there is room for more fools in the shaded area, who do not, however, play football. Which enclosure lies within which is the issue. The reader is asked to label the boundaries in the two diagrams.

Statement 1

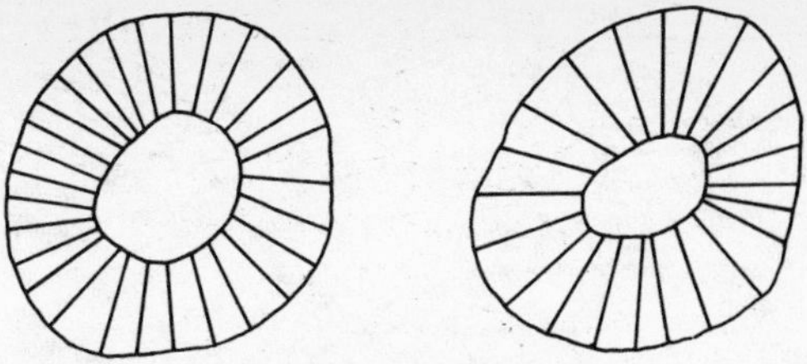

Statement 2

Footballing fools

In the Appendix we offer some more statements for you (p. 241).

There are just four pictures if there are two connected ideas. Our next diagram shows them, and asks that you allocate a picture to each of the pairs listed. They are discussed in the Appendix (pp. 241–2).

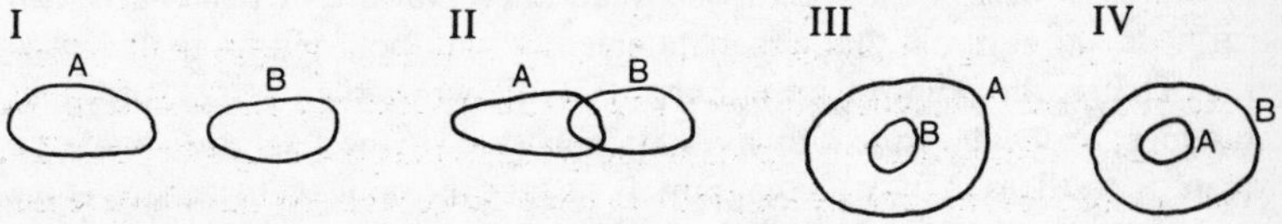

Assign a Roman number appropriately to each of the following pairs

(1) A Verbal
B Written
(2) A Even number
B Prime number
(3) A Footballers
B Cricketers
(4) A Square numbers
B Cube numbers
(5) A Square
B Rectangle
(6) A Value
B Price
(7) A Mathematics
B Calculation
(8) A Vowels
B Consonants
(9) A Line
B Point
(10) A Daffodils
B Flowers
(11) 'Top People take The Times'
(make your own sets)
(12) 'Cider is Bulmers'
(make your own sets)

These diagrams are not essential to thinking; some of us are happy and competent in language without visual aids; it is a matter of the way one's mind works. Yet these four images can prove a valuable mind resource for some of us. So many different pieces of information can be organized in this way.

This is a recurrent theme. Mathematics abstracts the relationships that exist; it removes the issue of what we are talking about, but is concerned with one idea lying within another, overlapping it, or being entirely distinct, as in our pictures. We are now better able to understand Russell's statement. We do not know what we are talking about, for the nature of the sets is not the issue.

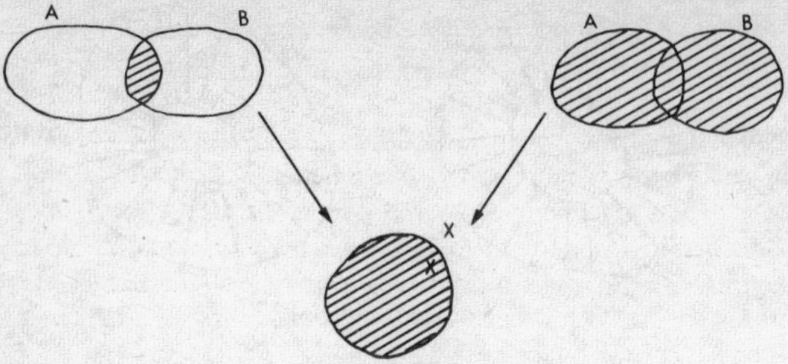

Cups and caps

It is a matter of how they relate one to another. Nor are we concerned with truth; we did not hold that either statement about fools and footballers was true. That is the point of his aphorism. Boole wrote the overlap, the 'and' statement, in an algebraic form as $A \cap B$, illustrated in the diagram. The 'or' he wrote as $A \cup B$, as in the next diagram. If we take the same set twice, we get the next picture, where cup and cap are the same area and we get $X^2 = X$. An odd result.

Lewis Carroll was not only a fellow of most delightful whimsy, but as C. H. Dodgson a mathematics don and a Fellow of Christ Church, Oxford. It is said that Queen Victoria, being amused by the Alice stories, urged Lewis Carroll to send her a copy of his next book. As C. H. Dodgson he sent her a treatise on algebra; we imagine she was not amused.

The theme of mathematical logic runs through his stories, and is again presented in a number of syllogisms, some of considerable complexity, which enlarge points already made. Syllogisms for long represented what logic was about; they date from Greek times and carry the imprimatur of Aristotle, who has good claims to be the originator of logic. It was he who asserted that a proposition and its contradiction could not co-exist, which led to the style of proof showing that there were infinitely many primes, which we gave in our first chapter. He also stated that a proposition was either true or false, known as the Law of the Excluded Middle, which we use in our syllogisms, even though there are some reservations about it.

The classical statement in logic is

'All men are mortal.'
'Socrates is a man.'

From these two statements we draw the conclusion

'Socrates is mortal.'

While this is no great exercise of the mind, it contains issues. The process is the point, not whether or not we accept the statements. If we regard Socrates as immortal it does not alter the validity of the reasoning.

The Venn diagrams emphasize what is already obvious. The two separate diagrams below represent the two statements. The visual display almost compels us to combine them to give the third, from which we readily draw the conclusion we already know. The detachment of meaning from process is a rather sophisticated notion in argument, not always understood.

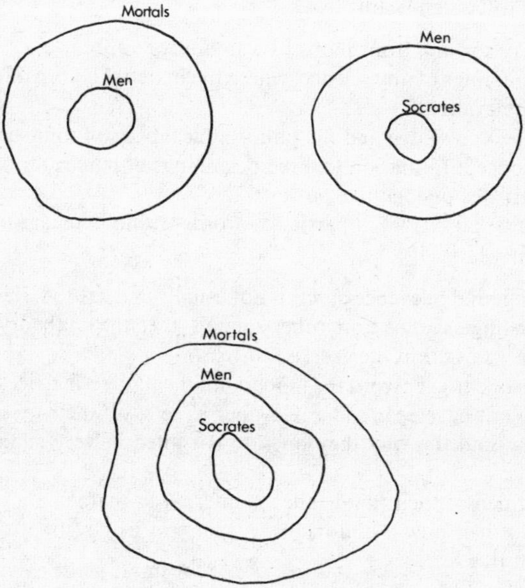

Socrates a mortal man

However linguistically competent we are, there is a level of complexity which the mind cannot handle, and that a symbolism will make routine. We shall not advance far in this direction but try to indicate what may be. Syllogisms may be extended to a string of statements. Lewis Carroll offered as many as ten from which we are expected to draw a single conclusion. We shall content ourselves with three:

(1) Ducks do not waltz.
(2) No officer ever declines to waltz.
(3) My poultry are all ducks.

It is not particularly profitable to discuss the truth or falsity of these statements. Process leads us to the conclusion

'My poultry are not officers.'

We discuss the detail in the Appendix (pp. 242–3).

Finally, of the issues concerning sets, a brief word about computers. The words we have been talking about, the logic words, are essential not only in the structure of our language, but in the way we address our computers. All forms of circuitry use Boolean algebra.

Let us list what we have said about sets, reflecting upon the areas where they appear to be significant:

(1) Everyday sorting, from shops to warehouses to libraries.
(2) The beginnings of any science, where the material is sorted into patterns to find relevance.
(3) Objective knowledge and the philosophical basis of number.
(4) Early concept formation and the beginnings of language.
(5) Comparisons of cognitive power.
(6) The logic that resides in particular words within language.
(7) Computer language.

The range and relevance of sets is not usually understood. If it is language that most distinguishes us from other animals, then the set theory that resides within it marks us off in the exercise of reason.

'Sets' is one trigger word for modern mathematics, 'matrices' is another. Matrices do not have as broad a relevance as sets, yet they are significant in mathematics, and the way they operate is rooted in very simple everyday matters.

Since much of their importance is internal to mathematics, let us start there. Even if it is now a dim memory, many of us have met simultaneous equations of this sort:

$$\begin{aligned} 5x-7y &= 3 \\ 3x+2y &= 8 \end{aligned}$$

There are various ways of going about solving them – multiplying both by something or other and adding or subtracting to get rid of one 'unknown', or drawing some graphs, or getting x 'in terms of' y from one and substituting in the other – or just plain guessing. If you end up with $x=2$ and $y=1$ then you have done the process correctly – even if it was guessing.

It was Cayley who first revamped such equations into a different form, suggested by the pattern of their layout. Rewrite them in the form

$$\begin{pmatrix} 5 & -7 \\ 3 & 2 \end{pmatrix}\begin{pmatrix} x \\ y \end{pmatrix} = \begin{pmatrix} 3 \\ 8 \end{pmatrix}$$

and we have moved into matrices. Technically all three of these 'arrays' are matrices. In fact if we wrote down any rectangular array of numbers, say

$$\begin{matrix} 2 & 3 & -4 & 8 \\ 7 & 2 & 1 & -3 \\ -8 & 2 & 4 & -7 \end{matrix}$$

it would be a matrix.

In mathematics they have a use when complicated groups of equations have to be solved, and more directly, when we use one set of unknowns in a number of equations as the bases for the next set of equations; but it is all a little complicated.

None the less, the way matrices work can be demonstrated in an everyday manner, even if they are not regularly used by the ordinary person. Suppose a chain of stores sell 'tweeters' and 'woofers' and in a week the sales are as follows:

	Tweeters	Woofers
High Street	15	10
Market Place	8	6
Downtown	7	5
Station Road	18	11
Centre Place	9	6

The array $\begin{pmatrix} 15 & 10 \\ 8 & 6 \\ 7 & 5 \\ 18 & 11 \\ 9 & 6 \end{pmatrix}$ is a matrix

As with a 'set' we start with a very general definition. Any such rectangular array is a matrix, whether we know what it means or not. Within the store we simply retain the numbers in this form (perhaps on a computer), and we know what the rows and columns stand for.

Suppose now that the next week is overall a little slacker in sales and the matrix turns out like this:

$$\begin{pmatrix} 11 & 9 \\ 7 & 7 \\ 5 & 3 \\ 14 & 12 \\ 6 & 3 \end{pmatrix}$$

We readily gather the meaning, without the knowledge of the headings. We now ask what the sales were in the two weeks. The rule of addition is obvious:

$$\begin{matrix}\text{Week 1} & & \text{Week 2} & & \text{Fortnight}\\ \begin{pmatrix}15 & 10\\ 8 & 6\\ 7 & 5\\ 18 & 11\\ 9 & 6\end{pmatrix} & + & \begin{pmatrix}11 & 9\\ 7 & 7\\ 5 & 3\\ 14 & 12\\ 6 & 3\end{pmatrix} & = & \begin{pmatrix}26 & 19\\ 15 & 13\\ 12 & 8\\ 32 & 23\\ 15 & 9\end{pmatrix}\end{matrix}$$

We have simply added corresponding numbers, or elements. This is a general rule for adding matrices, and we use it whatever the numbers mean, and whether we know what they mean or not. In one sense, mathematics does not seek to be tested in the real world; none the less it would fit where it touches. The rule for adding matrices is what we would expect from our example of shop sales.

In sets, we saw the union and intersection had analogies with adding and multiplication, though some of the rules came out slightly differently. With matrices, too, we need to look at multiplication as well as addition. The rule could be presented arbitrarily, but again there is a real life situation that can be presented in a matrix form. Back to our chain store. Shops are concerned about profit, not just about number of sales. Suppose they sell at a price to allow £20 profit on a 'tweeter' and £30 profit on a 'woofer'. We can handle the question of how much profit the first shop makes. It is 15 × £20 on tweeters and 10 × £30 on woofers. That is £300 + £300 or £600. Nothing remarkable in that. Work through shop by shop and we get the following table:

High Street	15 × 20 + 10 × 30	600
Market Place	8 × 20 + 6 × 30	340
Downtown	7 × 20 + 5 × 30	290
Station Road	18 × 20 + 11 × 30	690
Centre Place	9 × 20 + 6 × 30	360

Tabulation is always helpful. Being organized makes you feel in control. Look at the way we have written it and do some small changes. In particular, keep the matrix of sales we had started with

$$\begin{pmatrix}15 & 10\\ 8 & 6\\ 7 & 5\\ 18 & 11\\ 9 & 6\end{pmatrix} \times \begin{pmatrix}20\\ 30\end{pmatrix} = \begin{pmatrix}600\\ 340\\ 290\\ 690\\ 360\end{pmatrix}$$

The first and last matrices occur very naturally, but we need to justify the multiplication sign and the middle matrix. As far as

$$\begin{pmatrix}20\\ 30\end{pmatrix}$$

is concerned, it does record and store the information about profit per article. One aspect of matrices is that they are information stores. In some senses, writing it vertically rather than horizontally is a matter of convention. It is reasonable that the £20 and £30 should be written in this particular matrix form.

As for the multiplication sign, simply look at the meaning. The first matrix is goods sold, the second profit per article and third total profit. The operation must be multiplication. So by totally rational means we arrive at a way of multiplying matrices that looks most odd. We run along the rows of the first – such as 15 or 10, multiplying elements in the column of the second one, and adding the results. This is the sort of rule in mathematics that sounds complicated and can leave a sense of bewilderment. Why do mathematicians make such strange rules? They do so because, viewed properly, the rules are natural, not artificial. It is much easier to remember the shops selling tweeters and woofers and then recalling the rule we had to use (or simply rediscovering it) than to commit to rote memory the rule in the abstract. It is important to realize that we cannot add or multiply matrices unless they match in certain ways, coming out of our examples here.

In the first, when we added the matrices they were each the same shape and size. They must be, or addition is not possible. For multiplying, the number of 'elements' in the row of the first must equal the number in a column of the second.

We will do one addition and one multiplication in pure matrices, not knowing what the numbers stand for:

$$\begin{pmatrix} 1 & 3 & -1 & 8 \\ 2 & -4 & -7 & 3 \end{pmatrix} + \begin{pmatrix} 8 & 4 & -1 & -6 \\ -2 & 3 & -2 & -3 \end{pmatrix} = \begin{pmatrix} 9 & 7 & -2 & 2 \\ 0 & -1 & -9 & 0 \end{pmatrix}$$

We add corresponding elements; but there must be corresponding elements to add.

$$\begin{pmatrix} 1 & 3 & 2 \\ 2 & -1 & 4 \end{pmatrix} \begin{pmatrix} 2 & 1 \\ 3 & -2 \\ 4 & 1 \end{pmatrix} = \begin{pmatrix} 2+9+8 & 1-6+2 \\ 4-3+16 & 2+2+4 \end{pmatrix} = \begin{pmatrix} 19 & -3 \\ 17 & 8 \end{pmatrix}$$

As in algebra, we often leave out the multiply sign. The strangeness and mystique of mathematics is often dispelled by a demonstration that some rule or another does relate somewhere in the real world to a practical problem.

Though it is less easy to fit matrices into the real world than sets, other examples, such as routes on a map or electrical circuit theory, can be expressed in matrix form. One more connection is worth looking at. Take two very simple matrices and multiply them:

$$\begin{pmatrix} -1 & 0 \\ 0 & +1 \end{pmatrix} \begin{pmatrix} 2 \\ 3 \end{pmatrix} = \begin{pmatrix} -2 \\ +3 \end{pmatrix}$$

Straightforward enough, once we are comfortable with our rule for multiplication. Now we interpret them differently. Suppose the second matrix were not a profit per article, but the co-ordinates of a point, as shown in our diagram. The effect of the matrix

$$\begin{pmatrix} -1 & 0 \\ 0 & +1 \end{pmatrix}$$

is to move it to its new position below the x axis. It would do the same for any collection of points.

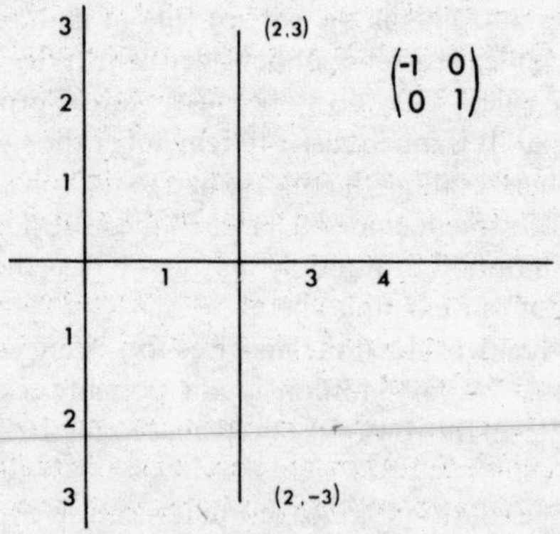

Transformations

In the next diagram we see how a number of points, and indeed a whole area, move under the action of the matrix. We get a 'reflection' in the x axis.

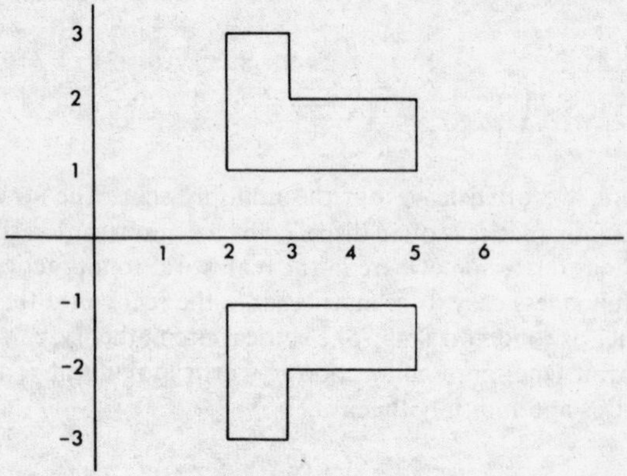

Reflections

That is the action of one particular simple matrix. Others may produce more complex shifts. It might be interesting to explore the effect of

$$\begin{pmatrix} 0 & +1 \\ +1 & 0 \end{pmatrix} \text{or} \begin{pmatrix} 0 & -1 \\ 1 & 0 \end{pmatrix}$$

or even

$$\begin{pmatrix} 1 & 0 \\ 0 & 1 \end{pmatrix}$$

It can at least be an amusing game.

The use of matrices in effecting changes of this sort (transformations) is not without point. There are problems, such as airflow past an aircraft wing or water past the supports of a bridge, that are quite complex if the shape is not simple. Transformations can be used in such problems to reduce the difficult solution to an easier one. It can be quite sophisticated, but the principle is to transform the problem to one that can be tackled, and a matrix may be one of the techniques for doing so.

In fact, the ramifications of matrices within mathematics are considerable, and some have very practical applications. Most of these lie beyond the everyday and in more technical areas.

We have not exhausted modern mathematics; no one could. The aim is simply to convince people that there are areas of mathematics developed only in the last century or two and not previously known. Modern mathematics is often very different from the common perception of mathematics as being concerned with more and more complicated calculations. Those are left nowadays to the computers.

III

Working with mathematics

11

Mathematics in action

At one time a sharp distinction was made between 'pure' mathematics and 'applied' mathematics. The very terms reflected an attitude well expressed in a toast by a Cambridge don at the turn of the century:

'To Pure Mathematics; and may it never be of use to anyone.'

People may have the opinions they choose, but they should not expect their hopes to be met. For most of us, it is an intriguing extra pleasure to discover that a piece of mathematics may actually prove useful in real life. That is not to say it is a main criterion for studying that piece of mathematics; it is simply a bonus when mathematics studied for its own sake proves of practical value.

The Greeks enjoyed ivory towers, and this was made possible by their position as aristocrats in a slave-based society. They were powerful and original thinkers and the elegance of some of their proofs in numbers and in geometry will appeal for ever. There is a deep aesthetic delight in their economy of thought. Yet we have seen that they believed, from what they held to be entirely rational considerations, that a weight ten times as heavy as another would fall ten times as fast. It does not. A child could devise an experiment that showed this. The first problems are in pure mathematics, and the falling weight is applied. Even the purist of last century would have accepted that tests in applied mathematics could be conducted in the physical world. Experiments of this sort were not acceptable to Greek thought.

There is an ironic twist in the way some of their studies worked out. They believed the circle to be the perfect shape. From this they deduced that heavenly motion was always circular and devised a system of heavenly spheres (the perfect three dimensional shape). They were not in fact far off, though any deviation from perfection would have been anathema to them.

From the circle, they moved to the cone, though it is unlikely they had a physical model of a cone; it was all in the mind. It might have proved a serious social gaffe to bring along a model to a group of Greek thinkers. If you slice a cone by a plane parallel to its base, at right angles to its axis, you get a circle, as we see in the diagram. If we cut elsewhere, again at right angles to the axis, we simply get smaller or larger circles.

A slice like this is called a 'section' and the Greeks decided to study 'conic sections'. Having sorted out a vast number of theorems about circles and triangles, they were looking for something more demanding.

If you tilt the plane of section, the shape you get is what we normally call 'oval'. The mathematical term for the shape is an ellipse. Since we have had to attempt perspective drawing, it looks remarkably like the circle on the base of

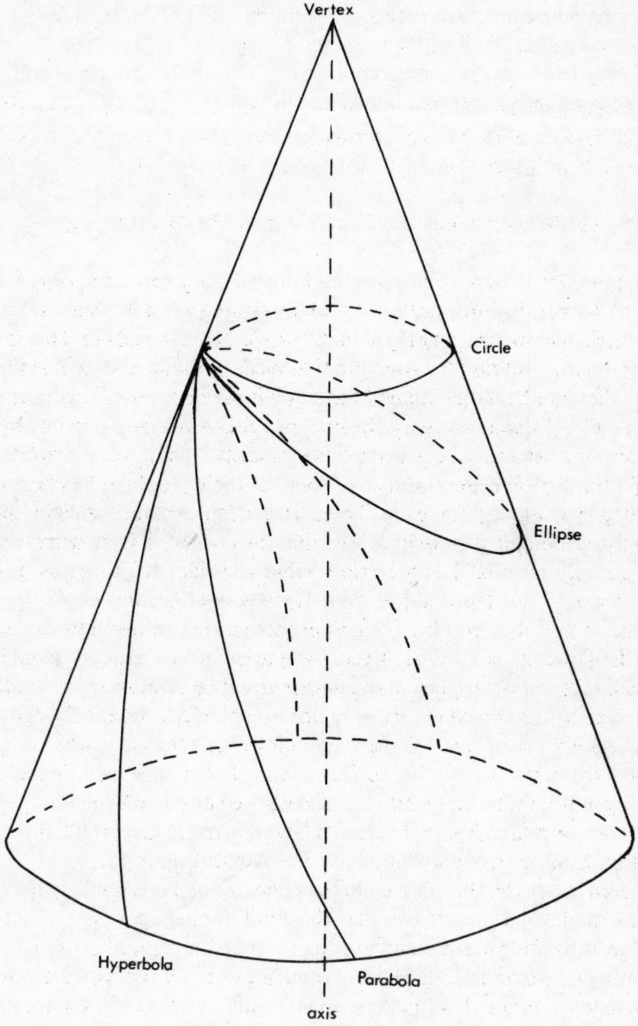

The conic sections

the cone. This is not surprising, for if you look at a circle from such an angle, it will appear elliptical.

If the plane tilts further we get a special case when it runs parallel to the 'edge' of the cone. They called it the 'parabola', and it will have a particular interest for us. Finally they tilted it yet further and got the hyperbola.

All these curves are pleasing to look at. We can understand why the Greeks believed the circle to be perfect, but it may be a more sophisticated aesthetic response that the other conic sections evoke. They obviously form a family closely associated with the circle.

Many facts were discovered about them. One of the most interesting was that the curves could be traced out by a point moving so that its distance from

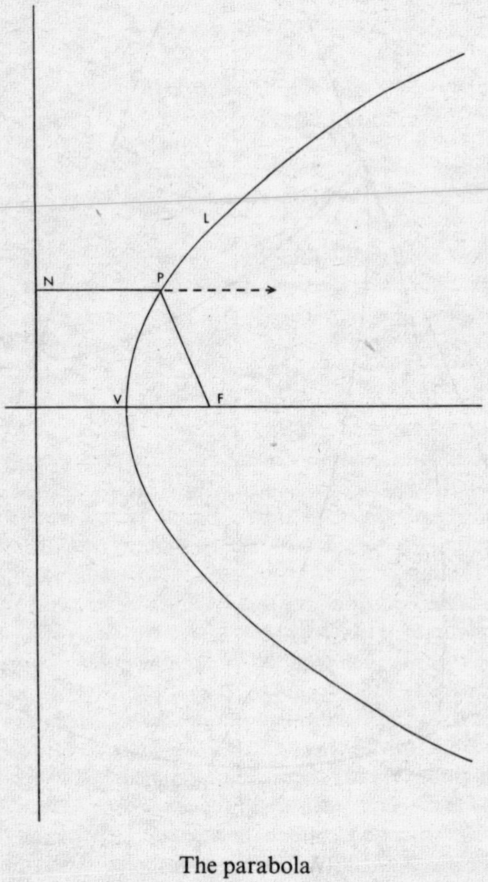

The parabola

a fixed line and its distance from a fixed point stayed fixed in proportion. That sounds quite complicated, and is easiest to see in the parabola.

In the diagram, the moving point is labelled P. The fixed point is called the focus and the fixed line (over at the left) the directrix. P is the same distance from both of them (PF = PN). Now let it move, but under the constraint that the distances remain equal.

It will go through V, where it is midway between the focus and the directrix, and L, where it is eight units from each. If we take a line, say twelve units out from the directrix, and find where it cuts the parabola, that point will be twelve units from the focus. It is worth taking some little while to familiarize ourselves with this property.

Here we have made the lengths equal, which gives us a parabola. If the point moved so that it was, say, always twice as far from the directrix as focus, we would get an ellipse. If it were (constantly) nearer the line than the point, we would get an even more open curve than the parabola – the hyperbola.

Should anyone doubt the power of those early mathematicians, they should themselves try to establish what is only stated here: the shapes we get by cutting the cone and the shapes we get by allowing a point to move under these rules are in fact the same shapes. The Greeks discovered what is called the 'focus–directrix' property, taking the curves from the cut cone. That is not an easy task.

This chapter is headed 'Mathematics in Action', yet we seem to be involved in some very abstract issues. In the real world the need to slice across a cone and examine the section does not often arise, even for ice-cream vendors. But we have some surprises in store, and the promised twist to the story.

Turn the parabola so the V becomes its topmost point. The curve is graceful and has a 'rightness' about it. It is also the path that a stone takes when it is thrown. This is a very practical issue. When studied by Galileo and others, it enabled the gunners of the time to hit their targets with considerable accuracy. The desire of men to lob large weights on to other men is not studied within mathematics; nor was the mathematics originally studied with this aim in view (at least in this case). Whatever moral judgement we might take, no one can deny that there was an exceedingly practical application of an originally abstract piece of knowledge.

The geometry of the parabola is pure mathematics. Dropping cannon balls on people is applied mathematics. The study of applied mathematics or mechanics is the study of moving or static objects subjected to various forces; we shall not discuss it further here, for we have wider fields to conquer.

With the development of electric light, the problem of focussing into a beam arose. To light a room, we are happy that the light from the filament in the bulb spreads itself in all directions. That is not appropriate for a torch, a car headlight, or a searchlight. We need to reflect the light from mirrors to send it in parallel straight lines – or as nearly parallel as we can. The solution is very amusing.

Return to the parabola. Place it back now with V at the left and the diagram upright. Simply place the source of light at the focus and reflect it with a parabola, treating it as a mirror. The cross-section of a car headlight mirror is exactly such a parabola. Every ray of light that leaves the focus bounces off the mirror and then heads off parallel to the axis. That is a fact. Proving it is quite another matter. It depends on a theorem, known to the Greeks, that if we draw a tangent to the curve at P (we have not drawn it, for we do not wish to clutter the diagram) it would bisect the angle between PN and PF. The parabola is an interesting curve.

In primary schools there is an activity known as curve-stitching, designed to help children with number relationships. One important piece of number work is learning what pairs of numbers add up to 20. A practical way of doing this is to take a piece of cardboard, draw two lines at right angles, each with the numbers 1–20 on them and make holes through the cardboard at every number. Coloured wool is then threaded on a needle and those numbers which add up to twenty are joined by stitching through the holes. The diagram shows the effect, though it is less attractive than when done on card with wool. You will observe that every pair of numbers giving twenty are formed by lines, to

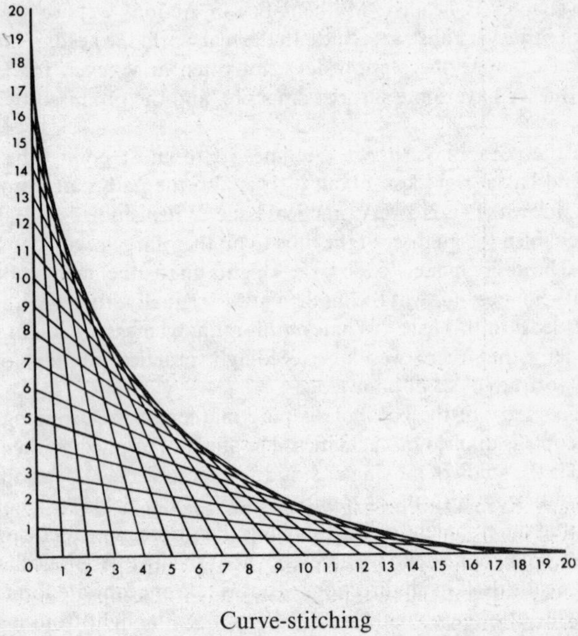

Curve-stitching

reinforce in the child's mind that, for example, 14+6, 7+13, 11+9, and so on add up to twenty.

The final effect is surprising. We have only drawn straight lines, and yet a curve has appeared. And the curve is a parabola. There is a 'resonance' here. The diagram we have drawn reminds us of some modern architecture, in particular some beautifully shaped roofs. (Remember the cars on the roof in the film *The Italian Job*?) It is not surprising that they do remind us of these. The roofs in question are usually hyperbolic paraboloids. That means that one way their sections are parabolas, and the other way hyperbolas. Ancient and modern are fused yet again.

We have mentioned the square numbers often and looked at the graph of $y=x^2$ in the calculus. The graph is a parabola. So too are various graphs we have already seen, such as $y=x^2-5x-6$. Indeed, if we only have x^2, xs and a number, we get a parabola when we draw it.

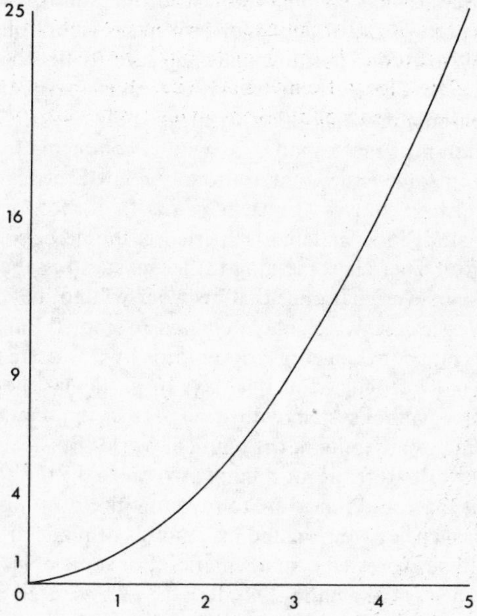

The squares

Perhaps the most surprising thing about the conic sections is that, after their analysis in ancient times, many centuries passed before they surfaced in a practical context (though this itself was several centuries before the search-light and curve stitching). It was not just in studying flight of cannon balls, but in a rather larger scheme.

The movement of celestial objects had been a source of deep interest to us through the ages, long before the importance of the sky in navigation stimulated more detailed study. Finally Kepler, with a vast number of observations and some monstrously heavy calculations, found three laws of planetary motion. For our purposes the interesting thing is that the planets travelled in ellipses, with the Sun as focus. The other laws established their speeds in their paths. Before long Newton had come up with his universal law of gravitation, which asserted that each body, Earth, Moon, Sun and so on, pulled on every other one with a force depending on their masses and the distances apart. Two pages of calculation, accessible to any undergraduate learning mathematics, now showed that the planets, comets, satellites all had to travel in one of the conic sections. The planets rotate in ellipses not far off circles. As we said, the Greeks were not very wrong. Of particular interest are comets. Some, of which the best known is Halley's, return at regular intervals. In fact it travels on a long flat ellipse, being visible only on the section of its track near the Sun. Others come in too fast for the Sun to hold. They move through on a hyperbola and disappear off into space. If for a planet or a comet we arranged its velocity and position with great care, it might travel in an exact circle or a parabola. These are the two special cases – normally it will be an ellipse or a hyperbola in which the body moves.

We started in pure mathematics, in a study conducted for its own sake, and we end up in some issues that are totally related to this world and the universe in which it moves. Einstein said: 'Is it not remarkable that mathematics, a study independent of experience, should be so adapted to the objects of reality?' There is something rather mysterious about it. No pure mathematician can ever guarantee that his work will not be of practical use.

Another classic case, which may well have prompted Einstein's observation, was the range of algebras developed by Grasseman, a brilliant mathematician not recognized in this day. In particular, he developed the tensor calculus, a symbol system with strange rules that could in no way be expected to relate to anything in the physical world. Einstein's observations on the physical world were the most innovative since the time of Newton. His speculations on space and time were to revolutionize our thinking about the universe. As was true of Newton and his analysis of planetary motion, these new ideas had to be expressed in mathematics. For some of his work, Newton had to create new mathematics. Modern physicists need very advanced mathematics as a tool to express their ideas. In general, like Einstein, they are not original mathematicians (by which we mean people who create new mathematics), whatever facility they have in its use. Einstein would not have regarded himself as a mathematician, despite his skill in using mathematics to express his ideas.

To explain his perception of space and time he needed some unusual mathematics, and he found it in the tensor calculus developed by Grasseman. The mathematics then predicts events in a physical world. Einstein's calculations showed that the planet Mercury would move a little differently from the way Newton had predicted as it passed through its nearest point to the Sun. Astronomers studied the path of Mercury at perihelion and found Einstein far nearer the truth then Newton. How, though, could we have predicted that this esoteric branch of mathematics, developed entirely within the bounds of mathematics and 'independent of experience', as Einstein has it, would be used to say where Mercury is? We are faced yet again with the question of the relationship between the patterns in our minds and those patterns we detect in (or do we project on to?) the external world.

There are other examples in mathematics where the work has been done before the need arises, and unconscious that there might be a need. In this century, however, the acceptance of what is mathematics has greatly extended, and in part this has been due to the deliberate attempts to develop mathematics to solve everyday problems. This is clearly the other way round from the examples of the conic sections and the tensor calculus that we have looked at. Some rather odd problems can lead into vast new areas of mathematics.

Suppose you are mowing the grass on a summer's evening and lose a valued ring from your finger. The sun is setting and you will feel very unsettled if you do not find it that evening. Think through what you might do. A person with any sense of purpose would not wander aimlessly around the lawn. Various policies might suggest themselves. An organized search, up and down in rows, preferably with others helping, might be a good idea. There might be reasons for searching more carefully in some parts than others; we might believe we lost it after a teabreak rather than before, and hence search the area mowed then rather than before the break. Try to think of other issues that might determine where we first looked. Altogether a rather mundane problem. Some might raise eyebrows that one should consider our policies in such detail in such a problem. Maybe so. Let us look at another problem.

During the last war, German submarines wrought havoc among the convoys of British shipping. This was not a minor issue. As in the First World War, it was one of the combat areas that might have proved decisive. There were two main defences to the submarine menace. One was the ship armed with radar and with depth charges. Another involved spotting from the air, and then attacking from the air. Only certain types of aircraft were adapted to this work and, as with all things, there were limits to what they could do.

Their range had limits, well known, clearly defined. They were limited in numbers, as were the pilots available to fly them. So it did not make sense simply to tell pilots to go out and look for submarines and make sure they had enough fuel to get back. Search policies were needed. Clearly there were certain ocean regions where submarines were most likely to be, determined by the convoy routes and the places where the submarines have to come from to reach them. If you know the convoy routes and the submarine bases it makes sense to spend more time looking there than elsewhere.

A skill of great importance in mathematics, and in many other studies, lies in seeing that problems are of the same type, that they have the same 'shape'. The term used in mathematics is 'isomorphic', which is only a word of Greek origin meaning 'having the same shape'. The important problem of the submarines and the more personal issue of the lost ring are isomorphic. Each involved searching with limited resources, against time pressures, over an area where some places are more likely than others. This is not what the average person thinks of as a mathematical problem. It was tackled with great analytical skill, in particular by J. D. Bernal, one of the most powerful scientific thinkers of this century. In dealing with this problem he created a new branch of mathematics known as operational research. Briefly it might be described as maximizing performance in certain systems under given constraints.

By an odd quirk there was a long lull in the development of operational research after the war, and Bernal went into crystallography. Furthermore, some of the work done was classified as secret. There is an amusing anecdote connected with this. Bernal, who was a well-known member of the communist party, wished some years after the war to refer to some work he did during the war and about which he wished to refresh his memory. He was refused on the grounds that he was a security risk.

The subject is now an important new branch of mathematics, and while some of the mathematics is very abstruse, some of the problems it tackles are easy to understand simply as problems, and give a good idea as to what the subject is about. Those chosen here are ones with a strong community aspect, and represent the sort of work that the Local Government Operation Research Unit based at Reading engages.

First, consider the bus service in a city. Policies in public transport are certainly a big issue, often with political implications. The interaction between the wishes of elected members of a local authority and the normal constraints in any system is a fascinating study in itself. At the simplest level, a council faced with a choice between alternatives A and B may have quite a marked preference for A. An analysis of the options may, however, reveal that A costs ten times as much as B. The council is then faced with the question, not whether they prefer A, but whether their preference is strong enough to override so great a difference in cost.

As far as a bus service is concerned, a local authority might ask for

(a) a better system at the same cost; or
(b) as good a system at a lesser cost.

The first task is to decide what they mean by 'good'. Judgemental terms of this sort may not always be easy to build into a mathematical system. Some behavioural studies must precede the mathematics, so either by direct observation or by means of surveys we try to find out what makes a system good in the eyes of the user. Suppose we find that, in general, people are unwilling to walk over a quarter of a mile to a bus stop, or to wait at it for more than a quarter of an hour. We have begun to quantify 'good'.

'Modelling' is an important technique with many problems in mathematics. In this case we try to reduce the service to its bare essentials, and end up with a network of lines, as we show in the diagram. We are used to doing this, in any case, with rail or underground diagrams. We need the lines, the points along them (bus stops) and the frequency of service along each route. Looking at that diagram we know it is a bus service, because we have said so, and labelled it thus. Present that diagram, unlabelled, to someone who has not read the text and they might well think it an electrical circuit or a rather unusual fish net.

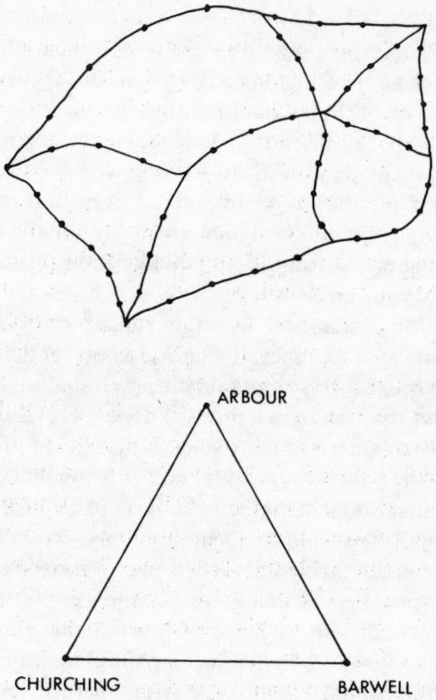

The bus service

Presumably now we know the cost per bus per mile. Because the problem is getting complicated, we feed it in a computer, with this data, and get it to cost the service. We now modify our diagram, by altering either the routes or the frequency. The computer can tell us how the cost varies; a human mind decides what modifications are sensible. We may need a separate program to tell us whether moving the routes causes more or less people to have to walk more than a quarter of a mile to a stop. The general approach may sound simple but decisions as to which modifications will most improve the service may rest on very sophisticated mathematics.

We started this chapter with a distinction between 'pure' and 'applied' mathematics. We might assume that our bus service problem was applied mathematics. In a sense this is true, but through historical chance the term 'applied' came to mean the study of forces acting upon bodies; matters such as the flight of a stone under gravity or the stresses in a roof rafter. Otherwise called mechanics, applied mathematics might more appropriately be thought of as a branch of physics. We have been left without a proper term for studies such as operational research and sometimes simply refer to the applications of mathematics.

Some problems in operational research lend themselves to very simple formulations which anyone might tackle. In doing so, there may be a degree of artificiality, but much of the flavour of the original issue remains. Look at our triangle diagram on p. 169. We now have three villages, conveniently placed at the corners of an equilateral triangle. Assume we have two buses each with two crews, that the number of inhabitants of Barwell is twice that of either Arbour or Churching, that the best pubs are in Arbour and the only cinema in Churching. Making what assumptions you like about the cost of running the buses, decide how you run the system.

Presented with a question like this, under the general heading of mathematics, many may feel uneasy. The 'looseness' of the statements will be worrying. This is because it is commonly believed that mathematics is coldly analytic and drives through to one precise answer. This is not the case. It can produce a number of answers, it can suggest that the answer lies in a certain range, it can vary the policies to produce different answers. Flexible questions demand flexible answers, and mathematics is capable of this.

Networks with flow along them are only one sort of work that operational research can tackle and further examples open up other areas. The availability of hospital beds is an important issue. People with uncomfortable or even disabling conditions such as arthritis may have to queue for a long time for a bed, on the grounds that however painful such an illness may be, it is not fatal. Yet the hospital has to cope with emergencies when people are rushed in following a heart attack or a road accident.

In consequence, however hard-pressed the health service may be, it is likely that there are empty beds in any hospital at any particular time. Someone who has been waiting years for a replacement hip might find it difficult to understand the system. This is one of a range of problems in 'queuing theory' which had developed into a significant body of knowledge within the applications of mathematics.

One approach is through 'simulation', and this can take the form of a game. The diagram indicates the way it is played. Two queues are formed, one urgent and one 'needy' but not urgent. Cards are drawn to build up these queues and different policies are adopted in reserving beds only for the urgent. The cards simply have numbers, indicating how many should be added to a queue; the numbers are overall in a statistical pattern. According to the restrictions placed on admissions from the 'needy' queue, at one extreme we cannot deal with the urgent, and at the other extreme we leave so many beds empty that the needy queue becomes excessive. Playing the game leads one to understand something of the problem involved for the hospital administration, and the existence of empty beds when people are waiting.

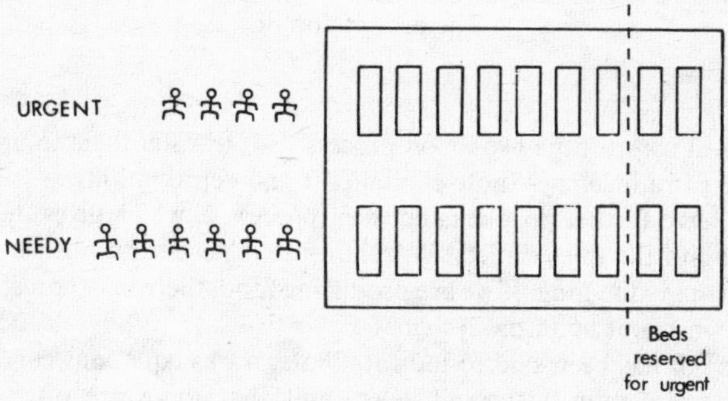

Waiting for a hospital bed

Simulation is of course a form of modelling. It is tempting to believe that all problems of an organizational nature can be solved by applying common sense. Apart from the fact that it it not so common, the evidence is that mathematical techniques can always do better than common sense in any issue beyond a certain level of complication. Yet the techniques are devised by the 'common sense' of those who originate them. A body of knowledge can simply be a repository of such refined common sense. We might say that this is what mathematics is.

A production line is a form of queuing. Our next diagram illustrates this. There are two processes to be performed, and people are employed to work at each. The raw material enters one end and is sold at the other as paper clips. At three stages we may need to store goods. It is commonplace to see large areas stocked with new cars ready to be sold near the major car producers.

One process may simply take longer than another. The solution to this may lie in getting more machines and people into process 1. If we get too many, the people in the second process cannot cope and we have a storage problem between the processes. Storage costs money even for paper clips. Not only that, but if the process continues the pile gets higher and higher. We may

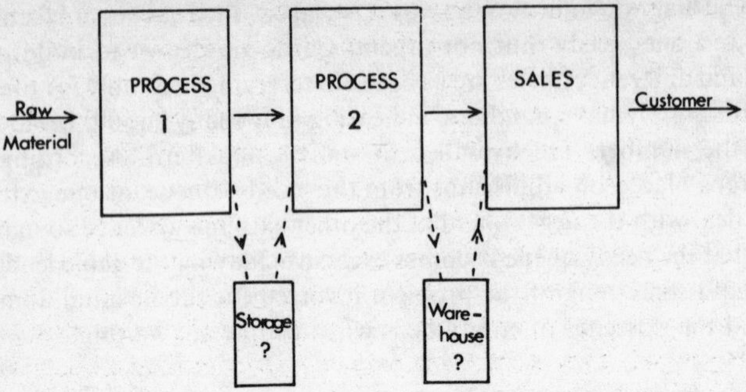

The production line

have to rest some of the people on process 1, or transfer them to process 2 if there are extra machines (note capital cost and depreciation).

We have further problems between process 2, which gives the finished product, and the sales outlets. Warehousing is necessary, but we must not keep too many too long. If we are producing food, there may be a store life, a further constraint upon the system.

Enough has been said to indicate that various equations can be set up between production costs and sales, and that these are often open to mathematical analysis. Even the local shopkeeper can get into difficulties if he is not sensible about stock control and his buying policies. He can often operate on his own common sense – and he learns from experience or goes broke. It is not suggested that he needs mathematical knowledge in this area to cope, but he is at the simpler end of vast production organizations where the techniques may not only be appropriate, they may be vital.

The mathematician is sometimes accused of treating people as numbers or as things. An individual mathematician may or may not do so. The view probably arises from the way in which he sees similarities between problems. The queues waiting to get into hospital and the queuing on the production line are the same sort of problem. You may well regard it as more important to enable the patient to have an operation quickly than to get the packets of paper clips into the stationers. Indeed it is. But the mathematician, in solving a general problem in queuing theory, seeks to achieve the best both for patient and the makers of the paper clips. He may treat both problems in the same way and with the same objectivity. It does not imply that he thinks one more important than the other. The worst situations are often achieved by muddled sentimentalists who are seized with the problem of one individual and allow them to jump the queue thereby causing more problems to others. Unrestrained sympathy, unrelieved by any rational process, is not the best way to run an organization which needs to be both sympathetic and fair.

Another category of problem is concerned with allocation. An issue that occasions much concern, and rightly so, is the allocation, within the state system, of children to secondary schools. Every year the mass media carry stories of children who cannot go to the schools they want, and whose parents consequently refuse to let the child go to the school to which he or she had been allocated. This is a serious human, social, and often political problem. It is worth looking at some of the policies adopted, their implications and the constraints within which the educational administration is obliged to work.

In some parts of the country a very simple policy is adopted, that of the catchment area. If you live within the catchment area of a school then you may go to it. In general, if you do not, you may only go to another school which has places remaining after all those from its catchment area have said they intend to go there or not. This style may be very appropriate for rural areas, where geographical considerations may dominate one's choice of school.

On the face of it, catchment areas are the simplest of policies, but look at our map of schools and the areas they serve. We have assumed it is largely urban and big enough to support five secondary schools. The major roads are an issue, although the youngest children may be over eleven years old. Catchment areas are indicated, but this arrangement clearly allows for what will certainly appear to parents to be anomalies. Parents with a house at H are nearer school B than school A, yet are in the latter's catchment area. If these parents strongly desire school B rather than A (and perhaps with good reason) the dictate of the local authority does not make sense.

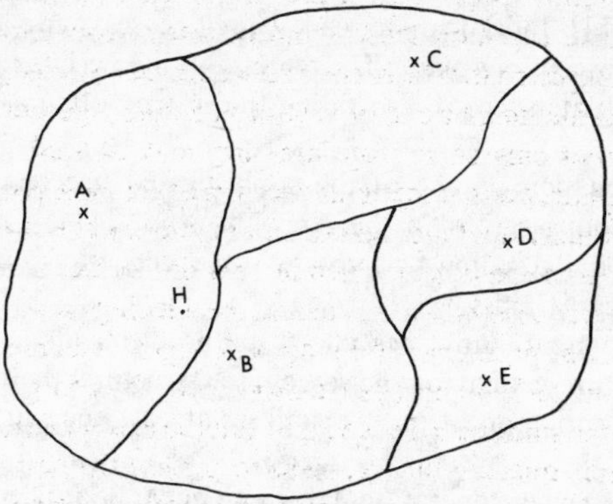

Catchment areas (1)

Now view it from the standpoint of the local authority. They have school buildings in positions which are seldom by design. Each one has a certain capacity, which should not be exceeded. They have determined the catchment area on the basis of what they know about the primary schools in the area, and hence the number of children transferring to secondary every year. They have arranged the catchment areas with due regard to this and to the major roads which may have to be crossed.

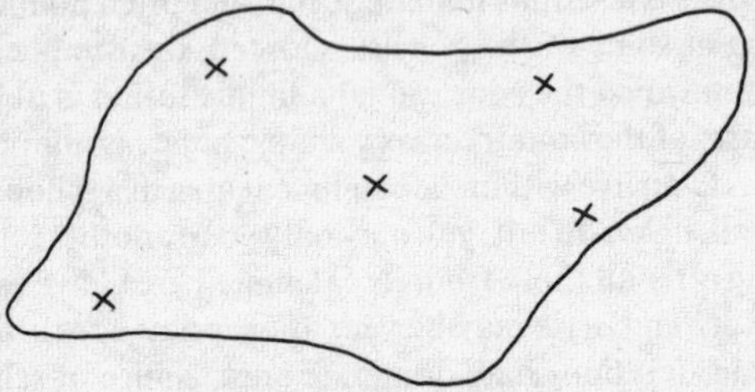

Catchment areas (2)

Both sides have essentially sensible positions. There is an interesting variant which those skilled in geometry may like to attempt. Draw an irregular area, populate it with, say, five secondary schools and then attempt to draw catchment areas such that everyone goes to the school nearest to where they live. To make the problem easier (even if unrealistic) assume a uniform population density over the region. It will reveal that it is easier to complain about bureaucracy than to solve the problems it faces.

But can there be no flexibility when someone has a strong desire to go to a particular school? Indeed there can. Single cases can, and often are, considered and exceptions made. The regrettable but perhaps inescapable fact is that the articulate succeed in bending the system, often at the expense of others.

A full discussion of one of the sophisticated processes of allocating children to school is given in the Appendix (pp. 243–5), together with a simplified model that allows us to do some allocation ourselves and to experience the constraints.

There are a great many such allocation problems all dealing with quite different situations, yet having the same basic structure; they are isomorphic to the pupil allocation problem. We shall see at least one when we later enter graph theory.

One last area where operational research might be applied – and it comes in two phases. The efficient use of 'plant' is something we have already touched upon, in the hospital and in manufacturing paper clips. Once we have established what physical resources are needed, there is the question of the way the processes in them are run.

You buy a large amount of space to set up a factory. Within any production system we need office space, shop floor, social amenities, and so on. The balance between them may depend heavily on the nature of the commercial or industrial enterprise – but getting the balance right matters. That is the first issue.

Next we must decide how we use the plant, who goes where and when and for how long. The distribution of facilities within a plant is a static problem; we then need to understand the dynamics of its use. In case this sounds rather high flown and detached, let us look at a genuine problem, not yet properly tackled, in the educational system.

If you were to walk around a large school, you might find nearly all the rooms in use, but many of them not used to capacity. Rooms capable of holding thirty people are occupied by remedial or sixth form groups of half a dozen students. It is not an economic use of space, but can we be sure of doing better? Certainly some of these thirty size classrooms might have been divided into three seminar rooms (perhaps some problems about the shape) thereby giving us two more 'teaching spaces'. The deputy head (whose task it is) might tell us that he or she is very tight on rooms – yet if the full classrooms had been replaced in some cases by smaller ones in the original building there might be much more freedom, with the same total square footage. It is not just a school problem, of course.

The school is quite complex in that teaching rooms are not the only provision. We have gymnasia, dining places, technical studies areas, drama workshops and so on. It mirrors the range of provision, though not the exact purposes, of a commercial business. A school building costs millions of public money; surely it makes sense that we build to meet the needs?

Some work has been done on this, but it is not an easy problem. Nor are we at the moment building many new schools, though we might well be adapting older ones to offer better balanced provision. In some areas, of declining rolls, it is not an issue. There is enough space for all activities. It does not alter the fact that space is being wasted.

The second phase, the 'process', in school terms means the timetable, and this is perhaps more complicated than in many other institutions. Blocks of students of different group sizes move through these resources every week. There are good and bad ways of arranging this. The analogy of pupils in commerce is of raw material being processed. We may not all see the educational process like this, but in the abstract mathematical model there is much in common.

Attempts on timetabling by mathematical processes, computer backed, have had a long history. It does not matter if the timetable is for a school or a railway, or if it is any sort of rota (or roster, as it is sometimes known). There is a process to be analysed and evolved to meet as many of the desired conditions as possible. Attempts on school timetabling (or scheduling as the Americans call it) seem to have originated in a major conference in Stockholm in 1962. We now have programs that are helpful, but can still leave the human timetabler with some knotty points at the end.

Once we can break the problem down into a series of routines (algorithms) an enormous amount of people's time may be saved. We shall now look at one problem not as difficult as the school timetable, but with enough factors to make it fairly complex.

Magistrates' courts vary greatly. Some may have several hundred justices of the peace, others only half a dozen. They serve very different areas and deal with vastly different problems. An interesting account of them is given in Elizabeth Burney's book *J.P.* Practices of who sits when and where may be varied, but there are some common factors which run through many benches.

Take a fairly typical bench with 75 magistrates. They have been appointed from the population at large through various routes, often through their political parties, but often also from personal contacts and a range of organizations. They are not paid, but deal with 97% of criminal offences. It is true that this includes all the less serious ones, and is dominated by traffic offences. Yet even a murderer must first appear before a justice, normally a lay person, unlearned in the law. Some of these justices may be retired people, and some may not have a full time job, but most do. They can only be expected to attend for half a day rather less than once a week. The Lord Chancellor lays down a minimum of 26 attendances per year, although it may run up to an average of over 40. Justices are often people of strong personality and organizing them is no mean task. Yet someone, normally one of the paid 'clerks', has the job of compiling a rota for when they sit. There are many human issues in this, and the clerks can be subject to strong pressures by individuals who do not want to sit on certain days or with certain other characters. In forming a rota we need to recognize (but perhaps not accept) that these pressures exist, and devise a rota, one kind of timetable, that satisfies a series of different and often conflicting criteria.

This occasions a great deal of work by many clerks in many magistrates courts over the whole country. It can take most of the time from October, when the composition of certain panels is decided, nearly to January, when the next year's rota needs to be ready to be sorted out. The mathematician seeks, not a way of sorting out next year's rota for a particular court, but a general way of going about rotas that short circuits much of the work and is available to all, irrespective of the size of the court or the nature of its work. A detailed discussion appears in the Appendix (pp. 245–8).

In reaching a solution of this problem, which can then be generally applied to courts anywhere, some set theory, some knowledge of primes and ordering patterns were necessary. However, the main issue is establishing an order of operations, an algorithm for the processes, and this is where thinking of an operational research type is used.

Let us look back and classify the types of issue to which we have sought to apply mathematics.

(1) Search patterns, be they looking for a lost ring or for submarines.

(2) Flow through networks; either a bus service or perhaps electrical circuitry.
(3) Queuing, either by patients for a hospital bed, or by paper clips to pass across a counter to a customer.
(4) Allocating pupils to secondary school places.
(5) Allocating resources appropriately to a production system.
(6) Programming, whether it be a school timetable or a magistrates' court rota.

Many are eased with efficient computer backing, but the essential thinking is human. It is a mathematical style of thinking and body of mathematical knowledge has grown to support the work, yet not everyone who enters this work was necessarily a mathematician by training, as is true of many of the best computer experts.

One of the newest branches of mathematics is graph theory (but not the sort of graphs we encountered in algebra). It might be baldly stated as the study of points joined by lines, which may seem a limited enough operation. Let us start as the subject starts, simple and with limited material. Our first problem may be stated thus: 'Six people gather together. Prove that there are either three who all know one another, or three who are all strangers.' We assume that if A knows B, then B knows A, and a similar situation for being strangers. A nice little logical problem, not quite as simple as it may seem. Some time spent on it before looking at the diagram and following through the solution would reveal that it is slightly more tricky than might at first appear.

The diagram promptly turns it into a problem about points and lines. The six points are the people and we may join them with two sorts of lines – say continuous and dotted. These, whichever way round, indicate either acquaintanceship or that the people at the ends of a line do not know one another. Effectively we have to show that however we do the joins (and all fifteen pairs must be joined), then we are bound to find a triangle with all its sides of the same sort.

We have made an abstract picture of the problem, draining it of social meaning, and looking at the mathematics. As we have already seen in

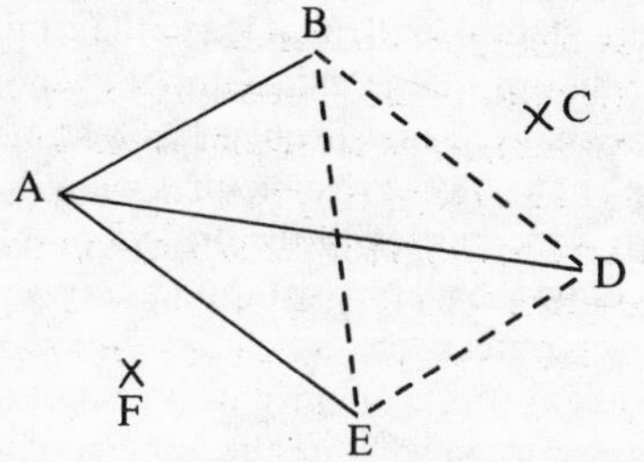

Friends and strangers

operational research, this process often reveals that problems that may look quite distinct in their social context are in fact exactly the same as far as their mathematical solution is concerned.

The reasoning goes thus. From A there are five lines joining the other points. There cannot help but be three of the same kind (they may of course be all of the same kind; but we did not specify the kind). In the diagram we have taken three and assumed they are continuous lines.

Now look at the three points they go to. These three form a triangle. If any one side of that triangle were a continuous line it would complete a triangle with two of the lines emanating from A. If not, then all three sides are dotted lines and we have the other sort of triangle. We cannot escape having one or the other.

The problem is amusing, not too deep, and may seem fairly trivial. Yet we do not always know what may turn out to be a starting point for greater things. Widely though it has been quoted, we cannot pass by Euler's solution of the bridges of Königsberg. The river flowing through Königsberg has two islands in it, the islands being connected to each other and the banks of the river by a total of seven bridges as in the diagram.

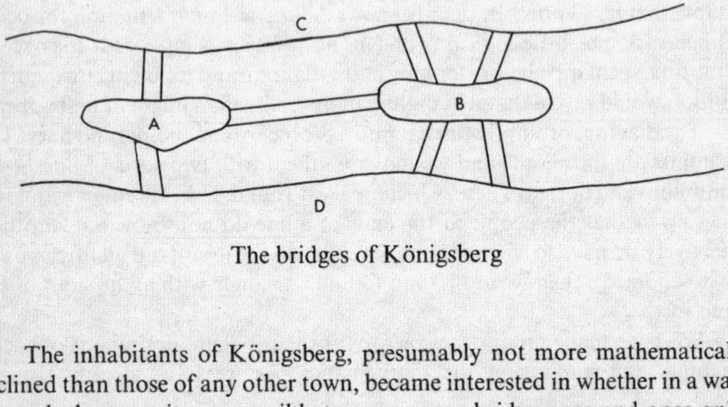

The bridges of Königsberg

The inhabitants of Königsberg, presumably not more mathematically inclined than those of any other town, became interested in whether in a walk through the town it was possible to cross every bridge once and once only. Again an amusing enough problem, but it led to the establishment by Euler of two interwoven branches of mathematics, topology and graph theory. Our second diagram presents the same problem as one of lines connecting up points. The reference points on the two pictures will help to establish that the graph we have drawn is identical with the bridges problem and that we now have to see if we can draw the picture without taking our pencil off the paper and without going over a line twice.

Trial and error might soon convince us it cannot be so drawn – but we need a proof. This hinges upon whether the number of lines running into the points is odd or even. If an even number run into the point you can visit and

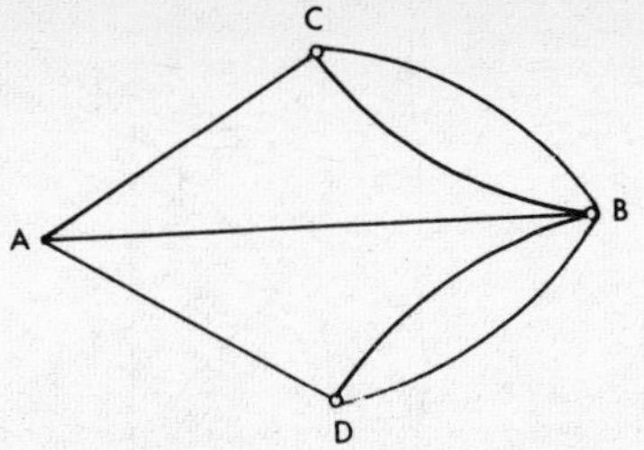

The bridges of Königsberg in diagrammatic form

leave, visit and leave and so on. If an odd number run in, then it must either be the starting point of our whole trip or the finish. If a second point has an odd number of lines, then it too must either be the start or finish. That is compatible. Bring in a third such vertex and we now cannot draw the picture in the manner demanded. In our picture, all four points have an odd number of lines running into them; and there is no way we can go for our suggested walk.

The interest in such a solution is that it not only serves this particular issue, but is a general method of attack on any similar problems. We simply check at each point whether an even or odd number of lines run into it, and if three or more come out odd we cannot do it.

A problem with a very long history runs as follows. 'A wolf, a goat and a cabbage are on one bank of a river. A ferryman wishes to take only one of them at a time. For obvious reasons, neither the wolf and the goat, nor the goat and the cabbage can be left unguarded. How is the ferryman to get them across the river?'

It is a fun problem. After playing about with it for a while most people can work it out; it is of manageable size and responds to a little organized thinking. Why has it anything to do with graph theory? Certainly we can represent the wolf, the goat and the cabbage by corners of a triangle, as we do in our diagram overleaf. We now draw firm lines on joins that are dangerous for one of the three and a dotted line on the other.

By now it is beginning to look like the acquaintance and stranger problem. The mathematician likes to unify his thinking. Ideally he would like to reduce all problems to one grand solution. It is now clear (and this could have been reached by many other paths) that the first move must be to take that corner of the triangle with two heavy lines over to the other bank. Only in this way is the dotted line in one bank. So we take the goat over. It would now be nice to 'flip' the triangle so that the dotted line is now on the other side. The three successive movements indicated (W, G, C) achieve this. As soon as the wolf arrives at the far side the goat goes back. Once the goat is back the

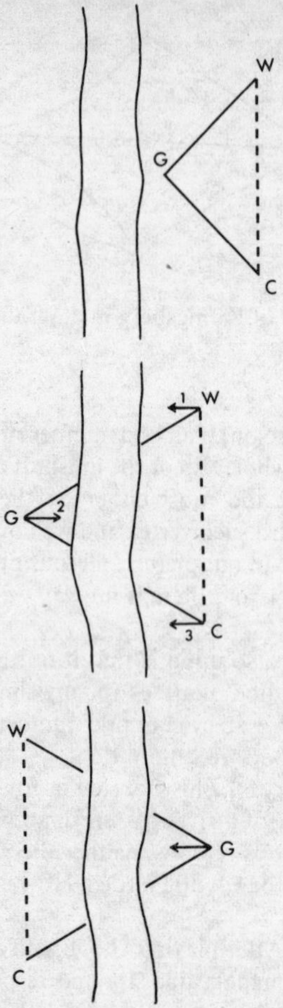

The river crossing

cabbage is removed. We have flipped the triangle over. We see that (C, G, W) would have done quite as well. Finally we retrieve the goat.

As with other problems it has been reduced to operations on mathematical objects – the points and lines of a triangle.

We will leave you with another very familiar puzzle, whose family relationship to the two we have discussed may again not be immediately obvious. 'Two men have a full 8 gallon jug of wine, and also two empty jugs of

5 and 3 gallon capacity respectively. What is the simplest way for them to divide the wine equally?' A solution appears in the Appendix (p. 249).

In introducing graph theory we have now discussed four problems easily understood and not in themselves deep, even if their study has led into more profound areas. Our next problem again starts in the real world and is one of the most famous within mathematics.

Early cartographers were led to wonder how many colours they needed for a map. It seems a simple starting point. The natural rule they had devised for drawing maps of countries, or regions within a country, was that if they put a boundary they coloured the regions on either side of it differently. We shall assume that a boundary is always a definite length of line, however it wriggles, and not consider point contact.

There was an economic issue, as there might be still in publishing today, of the number of colours we needed to use. Experience led them to the conclusion that four colours were always enough. If they started out wrongly they could find themselves in a situation where they needed a fifth, but by rethinking how they started it always seemed possible to do it with four.

This moves us out of the games and puzzles area into something of practical relevance, even if not of world-shattering importance. What it also did was to provide a problem that resisted the attempts on it of a number of major mathematicians for more than a century. If as an amateur you wish to make a name for yourself, we recommend either this problem or Goldbach's conjecture (see page 15).

The two diagrams show how we turn this problem into one of graph theory. We have simply taken a region which certainly needs four colours. There are four regions and six boundary lines. It may help to list the latter, using the areas they separate,

AB AC AD BC BD CD

To turn this into graph theory representation we, as it were, turn the whole thing inside out, so that the regions become points, while the lines separating regions become lines joining points. Our second diagram on p. 182 shows this. We have the same six lines listed above, and we now want the ends of each line to be of a different colour. It is the points we now think of as coloured.

Again we have an echo or a resonance. We seem to be asking that those joined by lines should be 'strangers'.

The four-colour conjecture has proved very difficult. The attempts made have resulted in much work being developed of considerable mathematical significance. We must take this on trust; it is not in the practical world that we mainly see the advantage. The cartographers needed no proof to draw their maps. Their experience and common sense allowed them to do so. It is only very recently that the four-colour problem has been solved, and in a manner that not all might find aesthetically pleasing. The problem was reduced to considering a large but not infinite number of possible configurations (that is,

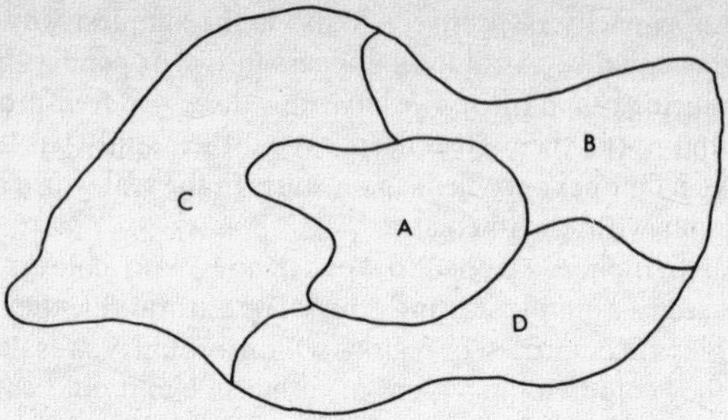

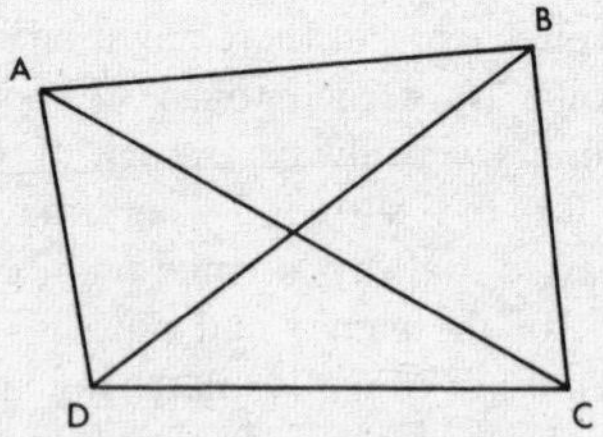

The 4-colour problem

numbers of points joined by lines). At this stage a computer tried all the possibilities and established effectively that four colours were always enough. We still hope that someone, one day, will 'see' why, in a manner easier for us all to accept.

This chapter purported to be about the applications of mathematics in the real world, and it is not perhaps impressive to sort out the bridges of Königsberg and how you colour a map. The emphasis has had to lie in problems accessible to our understanding, but the range of applications is in fact very considerable. We shall not just state this, but give a few examples, indicating the problems without providing the solutions, which are complex.

The main point being made is that many problems can be represented as points joined by lines. The planning of one-way systems in an urban area involves points joined directly by lines. One may get the impression that traffic planners seek solely to make certain points inaccessible, but that is not their intent, though it may be the result of their operations. The applications of mathematics here are certainly of highly practical relevance.

Perhaps two more problems, baldly stated, will suffice to show where mathematicians are active.

(1) In a certain factory there are a number of workers to be assigned to the same number of tasks. Every worker is capable of performing one or

more jobs. Can the workers necessarily be assigned one to each job satisfactorily?

(2) A company manufactures a number of chemicals. Some of these are incompatible and would cause explosions if brought into contact with each other. As a precautionary measure the company wishes to partition its warehouses into compartments, and store incompatible chemicals in different compartments. What is the least number of compartments into which the warehouse should be partitioned?

Let us now assess the role of mathematics in the real world. It impinges in various ways. Sometimes it has been developed for its own sake and is essentially 'pure' mathematics, but somehow it turns out to be useful in a variety of applications. This was so with the conic sections. Why this should happen is not clear, though we may speculate on the matters which our minds create, the physical world which we see, and the relationship between them. Our other examples, from operational research and graph theory (which obviously overlap to an extent), started as real problems and in some cases led to extremely abstruse and recondite mathematics not accessible to many. Mathematics appears to be independent of the problems to which it is applied. Both seem to exist in different worlds, though their interaction is evident. We need to explore these thoughts in another chapter.

12

Algorithms, problems and investigations

'When I use a word,' Humpty Dumpty said, in a rather scornful tone, 'it means just what I want it to mean – neither more nor less.' A very proper attitude.

The Shorter Oxford Dictionary tells us that the word 'algorithm' derives from 'Al Kuwarizmi', which means 'the man of Kuwarizmi'. This is the surname of the Arab mathematician Abu Ja'far Muhammad ibn Musa. It was his writings which introduced the Arabic system of numeration to Europe, and the word 'algorithm' then became attached to this system, and to the arithmetical processes that it made possible. Mathematics has many sources; we should not believe it to be a creation of Western civilization. In modern times the word algorithm has come to mean any routine process by which you proceed from a question to an answer. The emphasis is on the word 'routine'. It does not have to be an arithmetic process; it might lie in any area of mathematics, or might even be the set of rules by which you work out a rota. A computer cannot think, so it requires from you a set of rules, appropriate to its inner working, which it will then operate at great speed. It needs algorithms.

School mathematics in the past was dominated for many of us by the learning of algorithms, to the detriment of our enjoyment of mathematics and to the building of an assumption in so many that this is what mathematics is about. Algorithms have an important role, but we must not identify them with mathematical thought. Consider this:

'7987·569 ÷ 32·47 (ans. to 5 sig. fig.)'

Tedious and daunting. The part in brackets has a meaning, and if you do not remember it, a textbook will tell you. Let us look at the division itself. We can all remember, at least, that there is a method involving pencil and paper for doing this. The details may escape us, probably because despite assurances at school that such operations were a valuable tool for life, we have never used them. Nor have our lives been diminished by this fact.

Recapture the spirit of those days. Set out the long division in one of the approved manners.

(1) Move the decimal point in the divisor (32·47) to make it a number between 1 and 10. Thus 3·247.

(2) Move the decimal point in the dividend (7987·569) in the same direction, the same number of decimal places. Thus 798·7569. At this, you may feel a growing sense of boredom, tinged perhaps with a rising feeling of panic. This is not at all surprising. It is boring and causes anxiety. But bear with it the while.

(3) Set the numbers out as shown below:

$$3{\cdot}247\ \overline{)798{\cdot}7569}$$

(4) Find how often the 3 goes into the 7. (It is not really 7, it is 700, but no matter.) Try multiplying the whole divisor by this number.

(5) If it is below the dividend, write that number above the seven, and that number multiplied by the divisor below the dividend.

$$\begin{array}{rl} & 2 \\ 3{\cdot}247 & \overline{)798{\cdot}7569} \\ & 649{\cdot}4 \end{array}$$

Be careful where you put the 4 when you start multiplying.

Enough is enough; we shall not plague you further. An intelligent person can respond to a complicated series of rules like this, but common sense tells us that it cannot be a worthwhile activity. It is not acceptable to respond to rules without meaning.

The real result is that we are being treated as machines. That is the issue, and the resolution of it is now with us. We have machines. Machines more efficient, greatly more accurate, and many times faster than we are. Insert that division into your pocket calculator and in no time (and with no stress) we get

$$7987{\cdot}569 \div 32{\cdot}47 = 245{\cdot}99843$$

The decimal does not really stop there; but the calculator does. As for '5 sig. fig.' cut off after 5 figures like this

$$245{\cdot}99 \mid 848$$

and then correct up. We get 246·00. So the answer was 246, 'as near as damn it'. The only reason the machine can beat us with such ridiculous ease in arriving at this answer is that the original question had an algorithmic solution. The way in which we divide one number into another has been known for many centuries. It is codified, 'programmed', routinized.

That is not to diminish in any way the originator of the pencil and paper method by which we do long division. He, or she, was able in the extreme. The devising of algorithms is in quite a different category from the using of them.

Nor should we believe that there is no virtue in teaching long division. It is possible to show why it works, stage by stage. This is a fascinating insight into the difficulties experienced by the originator of the method. Yet the rule-based instructions with which we started are valueless. They instruct the dull, and repel the intelligent. They engage the conformer in a discipline in which if he continues to conform, he will fail, and they obscure from the creative the pleasure that they might derive from mathematics.

In extreme cases, the presentation of mathematics as being based upon arbitrary rules, whose reason may not be sought, can be psychologically damaging. A research subject was a highly intelligent and successful educationalist. She had a first class degree in English and an extremely successful career. Yet she felt crippled and greatly diminished by her incomprehension of mathematics. It started at the tables and long division was a dragon, not to be tackled. It was explained that long division meant taking one number away from another as often as possible, and keeping track of how often. It helped to work in bigger chunks, and take off, say, 30 times the number at one go, to cut down the work. In half an hour a thirty-year-old dragon was slain. Dramatic yet true.

Algorithms do have value. It would be foolish to suggest that it is not convenient to know paper and pencil techniques for adding a simple string of numbers. The greengrocer and butcher always prefer it to the computerized till. The danger is that people can be led to believe that all of mathematics consists of such processes, and in particular that these should be performed without understanding.

Algorithms do not have to be about calculation, but may relate to any series of routine procedures. Suppose someone devises a system for a rota. It may be a duty rota in a hospital or a railway, or our rota for magistrates sitting in a court. Anyone who has attempted the planning of any such timetable knows of the difficulties. There are usually many conflicting needs, not easily reconciled, and likely to lead to much annoyance if not met. At first sight they all seem individual and not amenable to any systematization. If they can be systematized (and it is not always possible) the system asks a string of questions, easy to answer, and then tells you what to do. Again, the person who has devised the routine may have engaged in a very demanding intellectual process. The user of the process may not even have to think very much. Indeed, the devisor may have that in mind as one of his criteria.

When such a matter can be reduced in this manner then we have an algorithm, a defined set of procedures: they may occur in subjects other than mathematics. It may be possible to define a series of tests to determine whether a piece of prose satisfies certain grammatical rules. It would not be easy or perhaps even possible to devise an algorithm that told you anything of the quality of the writing. Mathematics contains many algorithmic processes. These include the 'four rules' but naturally include the rules for solving various algebraic equations, and those for performing various geometric constructions. Nor is it only at the most elementary level that such rules are used. Certain processes in the differential and integral calculus have also been reduced to a series of rules. Algorithms exist at all levels. Yet to learn them is not to experience mathematics in any real sense. To enjoy a subject, more is needed. It may well be that the scientist or engineer sees mathematics simply as a tool (though most able ones would think it much more than that). They may get their satisfactions from perceptions within their own disciplines, and welcome the routines that mathematics can offer them. The satisfactions in mathematics lie in thinking mathematically in a creative sense, and this is not something open to only a few, a select priesthood of mathematicians, but is available to all. To understand this, we need to look beyond algorithms.

Many of us may remember from our schooldays a distinction in our textbooks between 'sums' and 'problems'. The most obvious distinction was that there were words in problems and not in sums. If the question reads 'A man buys six 5-litre tins of paint at £5.63 a tin, and then has to pay VAT at 15%; what is the total cost?' then we are clearly dealing with a sum, but it is somewhat disguised. The 'problem' lies in the language in which it is wrapped, rather than the mathematics. Let us unravel it, at least to the calculation stage, and then reflect upon it. The size of the tin does not matter since we are told its total cost; so we must beware of bringing the '5' from the '5-litre' into the calculation at all. The number of tins and their individual cost is important, and we need to realize that the mathematical process involved is multiplication. We could stop here and work out $6 \times 5{\cdot}63$, preferably on a calculator because it will be easier and more reliable. Answer 33·78. The next step is more sophisticated. We have to add on 15%. One way is to find 15% or 15/100 and then add it on. With the calculator it is easier simply to multiply by 1·15 (try to remember why). Answer 38·847. So we are probably asked for £38·85.

In solving this 'problem' we needed to discover the computations that had to be made. They were not told to us explicitly. The process is both more demanding and more useful. Yet it is not what a mathematician would classify as problem-solving. That is not because it is relatively easy. Problems can be anything from very easy to impossible. Whether something is a problem lies deeper than that. In reading about the tins of paint we needed to decide what algorithms we should use, and the clues as to which they were lay in the words. Having solved this task, however, we found no mathematical difficulties; routines were available. We can now move on to see what constitutes a problem.

As we said, problems come in all shapes and sizes. Not only may they vary greatly in difficulty, they may also be interesting and important mathematically or they may be what the mathematician calls 'trivial'. This is not as dismissive as it may sound. Often such a problem is easy (at least to a mathematician) but even that need not be so. In the mathematical sense 'trivial' means that it is not of significance and does not advance our knowledge.

Let us first look at a problem that is quite amusing, but essentially trivial. This most certainly does not mean that we should feel diminished if we cannot do it; it is likely that most people would not be able to do it, even though the mathematics needed is well within everyone's grasp. It is a genuine problem for most of us and the characteristics of a problem are that we know our starting point, we know our goal, but we do not have routine processes for getting there. There is an increasing interest in problem-solving as a method of learning, a welcome departure from the formal education many of us will have received. So consider this proposition:

'Apart from 2 and 3, every prime number lies next to a multiple of 6.'

The problem is to establish whether this is true or not. What makes it a problem is that there seems to be no special way of going about it.

There are ways of going about problem-solving, ways that can be taught. The first rule is not to start solving straight away. We could sloganize it thus:

Rule 1: 'DON'T START!'

Many might choose to take that very literally. What is meant is that no attempt should initially be made to reach a solution. Familiarize yourself with the problem. Make sure you know exactly what it means and what you are asked to do. Check back that you remember what a prime number is. Get clear what 'lies next to' means. Presumably it means one above or one below a number that divides exactly by 6. The statement says 'Apart from 2 and 3' and the reason must be that they certainly do not lie next to a multiple of 6, so our next statement is

Rule 2: 'STABILIZE THE PROBLEM'

It is easy to get into a muddle with things dropping out of our minds and our losing what it was that we had to do. Examine the problem from all angles – get it firmly established in the mind.

The word 'every' is a difficulty. Suppose it were not there. Could we tackle a similar problem where there was a restriction on the number of primes? Suppose we asked 'Do the primes from 5 to 29 all lie next to a multiple of 6?' We have now started towards a solution, but with another banner:

Rule 3: 'TRY AN EASY CASE'.

With this simple case, we have an algorithm, and a very simple one. Start with 5 and work up through the primes, looking at them:

5 is 6 minus 1; 7 is 6 plus 1; 11 is 12 minus 1; 13 is 12 plus 1; 17 is 18 minus 1; 19 is 18 plus 1; 23 is 24 minus 1; 29 is 30 minus 1

So it seems to work. Notice, of course, that there are examples of numbers next to multiples of 6 that are not primes. For instance 25 and 27. At this stage let us have another slogan:

Rule 4: 'REFLECT'

Think out where we are and what we know. We may be a bit clearer about the question. It says that primes are next door to multiples of 6, not that numbers next door to multiples of 6 are prime. Some certainly are not. It is surprising how easy it is to switch a statement round and think we are proving something other than we are. Remember our valid and invalid arguments in the last chapter but one. But reflecting shows us the beginning of a pattern. 5 and 7 lie around 6. 11 and 13 around 12, 17 and 19 around 18. We are beginning to get a feel for the problem. Another suggestion:

Rule 5: 'DRAW A DIAGRAM'

It is worth remarking at this stage that we may not always need to use all our slogans, nor will they always come in the same order. Later problems may throw up more slogans. The advantage of them is that when you do not know what to do, they may give a clue as to a new path we may explore.

Let us draw the diagram (No. 1).

(1)

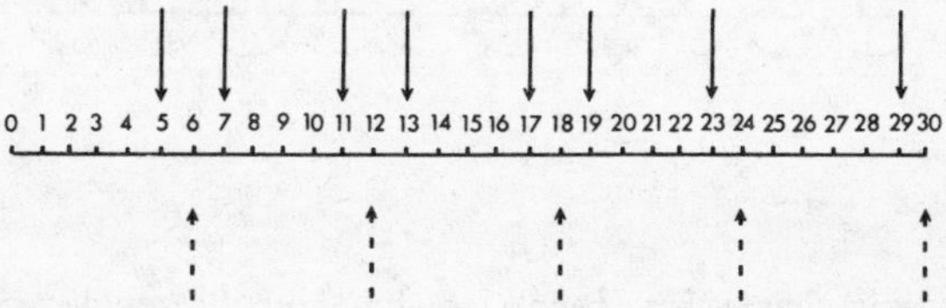

Pictures help. Some of us find them more useful than others, and not everyone sees things easily when presented spatially, but always at least consider supplementing the words with pictures.

Study the primes. Can we extend it to cover 'every' prime? What would the picture look like much higher up the number line? Diagram 2 shows a part of the line further up. This time we have marked where the multiples of 6 come, as we did before. Then we marked where the primes might come; we cannot tell where they *do* come unless we know whereabouts we are on a number line that goes on for ever.

(2)

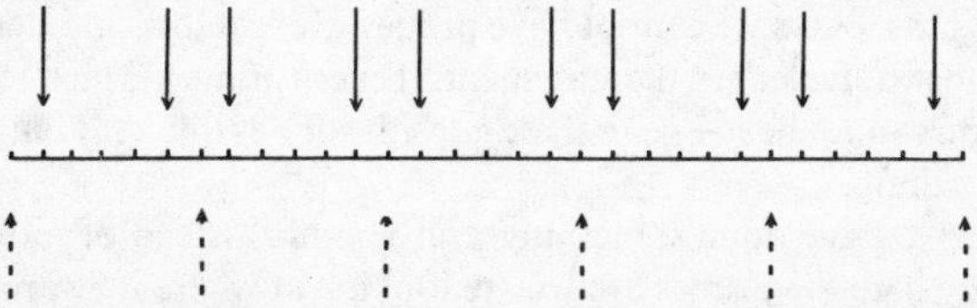

The oddest thing about this number line is that it has no numbers. There are marks where they will be, but we have no idea how high up we are. We have a general, not a specific, chunk of line. This is a sign for another piece of guidance in problem solving:

Rule 6: 'GENERALIZE'

The virtue in this problem is that it spoke of 'every' prime beyond 2 and 3 – and that means we must become general and not specific. The picture meets that need. If we can sort out something about that strip of line, it will apply everywhere.

There is another very important issue about problems involving the word 'every'. We have seen, when we took an easy case, that we were back with an algorithm. Check them out one by one. That would be very tedious if we had wanted, for instance, to look at the numbers up to 1000. If we had a microcomputer beside us, it would rattle it off in next to no time. That is a big step, but not enough. You cannot ask a computer to try 'every' prime. However fast it goes it will be on an endless treadmill. So we have to resort to thinking. Algorithms are devised to save us the need to think.

(3)

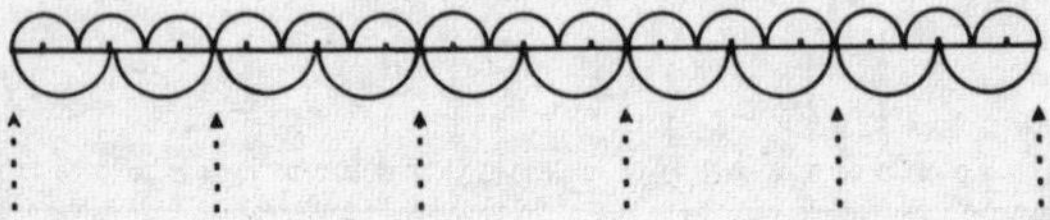

It may seem that we have been a long time with this problem already, but if in taking our time we learn something about the way in which we tackle problems, it will pay off in the long run. We now have this general line very neatly patterned. It is regular and not capricious, as is the way primes crop up. Perhaps we should list another idea:

Rule 7: 'SEARCH FOR PATTERN'

Pattern usually has a meaning attached to it. Here the pattern is steady and regular. There is a place where the proposition says we might find a prime, then a multiple of 6, then another place for a possible prime, then a gap of three places and the pattern repeats.

In the gaps it says we cannot have primes. Reflect for a moment to check that this is the equivalent to the statement. If every prime lies next to a multiple of 6 then you cannot have one in these gaps. It sounds different but it amounts to the same thing.

Again, we have done something a little odd. Instead of looking at the problem of where the primes are we are looking at where they are not. In set language we are looking at a 'complement' of the set we were considering. Do not let this technique escape us. It may help again later.

Rule 8: 'CONSIDER THE COMPLEMENT'

Look in the first diagram at the numbers in the gaps. In batches of three they are 8, 9, 10; 14, 15, 16; 20, 21, 22 and 26, 27, 28. Can we see a common pattern which explains why they are not prime? Certainly in every case the outer two are even. The middle one is odd, but it is divisible by three. We are well on the way in our thinking, but it is simply no use establishing just for numbers up to thirty. We need a general reason why these numbers have these divisors. So we move to diagram 2, where we had generalized. The reasons now stare us in the face.

The outer numbers of those '3-gaps' are always two steps away from a multiple of 6. Since 6 is even, so must any multiple of 6, and so also must numbers 2 away. Reflect. As for the middle number in each '3-gap', it is halfway between multiples of 6 and must therefore divide by three.

In diagram 3 we have retained the pointers for multiples of 6 and have hopped in 2s along the top of the line, marking all the even numbers, and in 3s along the bottom, marking multiples of 3. Naturally the 2 hops and the 3 hops keep meeting at the 6s. The only points which could possibly be prime, since numbers dividing by 2 or 3 are not, must be next door to a multiple of 6. We have solved the problem. We can also see more clearly why 2 or 3 were excluded. We needed to look at their multiples – but not themselves.

Offer this problem to a mathematician, and he might well see it immediately, and brush off the question. He thinks it 'trivial'. Technically that is so. The problem has taught us nothing about primes except that they cannot land on multiples of 2 or 3, which is obvious enough. Yet as a problem it is valuable as an illustration. A number of modes of thinking are involved and it demonstrates clearly what a problem is, for when we started there was no obvious routine which would lead us to an answer.

There is an abundance of problems of all sorts in number theory, and it is certainly not easy to judge how difficult they may prove. Mathematicians may claim to have intuitions about levels of difficulty, and there may be something of a paradox about operating with intuition in an area involving pure logic. It is, of course, an intuition based on long experience, retained but not explicit, of similar problems. However, it may be amusing to ask people these two questions:

> 'Can the gaps between two successive prime numbers be as large as you like?'
> 'Is there always a prime number between two successive square numbers?'

Firstly, ask for an intuitive assessment of level of difficulty. Then let an attempt be made on them. One of the two questions is quite easy to solve for anyone who has worked at this sort of problem before. It is not trivial, but is quite easy. The other is an unsolved problem, and there are many of these in mathematics. One of these two is solved in the Appendix – the easy one (pp. 249–50).

That there are unsolved problems is a surprise to many people, and it is a part of the nature of mathematics that everyone should know. There are many speculations which seem likely to be true, on the basis of those numbers we have tried, but for which no general proof has been established. This is a little game you can try on a pocket calculator. Take any number you like. If it is even, then divide it by 2; if it is odd, multiply by 3 and then add 1. Having done that, repeat the process with the number you have got.

Suppose we took 23

$$23 \rightarrow 70 \rightarrow 35 \rightarrow 106 \rightarrow 53 \rightarrow 160 \rightarrow 80 \rightarrow 40 \rightarrow 20 \rightarrow 10 \rightarrow 5 \rightarrow 16 \rightarrow 8 \rightarrow 4 \rightarrow 2 \rightarrow 1$$

We reach 1, after which it can go up to 4 again and back through 2 to 1 in an endless loop.

It is fun to play with, the strings can get very long, but we always seem to return to 1, no matter where we start. This is a numerical oddity, which may seem of passing interest. Yet mathematicians have tried, hard and unsuccessfully, to prove that it always works. Maybe it matters, maybe it does not; but it is certainly not easy. It is known as the Syracuse conjecture.

Another famous result, where it is equally easy to see what the problem is, is Goldbach's conjecture, mentioned in our first chapter (p. 15).

'Can every even number greater than 4 be expressed as two prime numbers added together?'

We do not know. It may be that we never shall. This problem has been found interesting, as well as difficult, by mathematicians. It is not altogether easy to say what is 'important' mathematically. It does not hinge on practical everyday usefulness, but rather on whether it is a problem related to a number of others that, if solved, might sort out quite a region of the subject. Think of mathematics as an enormous region of which only parts are cultivated, and there are people working furiously to bring the rest under some sort of control.

Though it is tempting to remain with problems about the ordinary (counting) numbers, there are problems, both piffling and profound, in other areas. Let us edge away slightly from numbers and look at an amusement with no claim to mathematical depth: the milk-crate problem.

'In a milk crate, with 6×4 compartments, there are 18 bottles. Can they be arranged so that there is an even number of bottles in every row and every column?'

One way would be to get a milk crate and some bottles and start playing around. Not, in fact, a bad idea. It would be a good way to stabilize the problem. But do any other of our slogans help if we are not so equipped? Certainly. Draw a diagram.

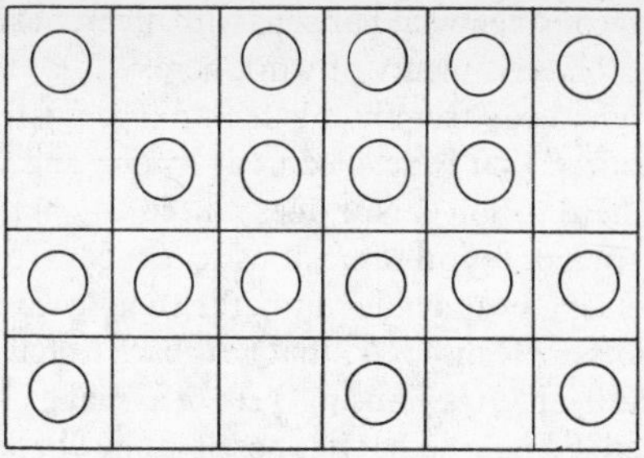

It may not look much like a crate, or bottles, but realistic pictures are not the point. A mathematical diagram contains simply the mathematical issues. We do not need to know if they are silver-topped or gold-topped bottles, and the diagram would be just as valuable for our purposes if they were Xs and not Os.

Are any other of our slogans useful? What about 'Try an easy case'? That might be worth while. Draw a crate 4 × 2 and use only 4 bottles. It would give a feel for the problem. There is a short circuit to this problem. Rule 7 suggests 'Consider the complement'. Here it is a very pretty device. In the earlier problem, instead of looking at where the primes were, we looked at where they were not. Different problem – same technique. We do not look at where the bottles are, but where they are not. There are six spaces. If we arrange those so that there is an even number in every row and column, then the same will be true of the bottles. A new diagram labels the spaces (this time with Xs). A little of the chess player's art and we make two moves as shown by the dotted lines and crosses, and reading column by column we have 2, 2, 0, 0, 0, 2. So there is an even number (zero is even for our purposes) in every row and column. Put the bottles in and our problem is solved.

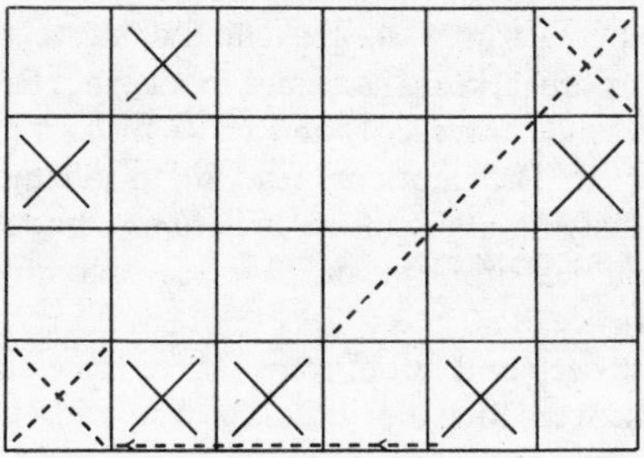

This is the sort of problem that lends itself to generalization. The numbers, 24 compartments and 18 bottles, were arbitrary. Mathematical thinking drives us to consider what happens with different size crates, different numbers of bottles. There is plenty of work here.

Let us move right away from numbers into a geometrical problem. Not one of the Euclid 'riders' that bedevilled us at school, and yet a problem easy to understand and hard to solve. Nor does it need any mathematics to speak of. An apparently trick question some of us remember from childhood went thus: 'A hunter pursues a bear a mile south, then a mile west and finally a mile north, when he shoots it. He discovers that he is back exactly where he started. What colour was the bear?' It is amusing, yet not a 'shaggy bear' story, despite its context. You need to find whether it is possible anywhere on the earth to do what it is claimed the hunter did. Can we go a mile south, a mile west and then a mile north and be back where we started? There is no trickery. North and south are directions along lines of longitude. East and west are plus and minus along lines of latitude.

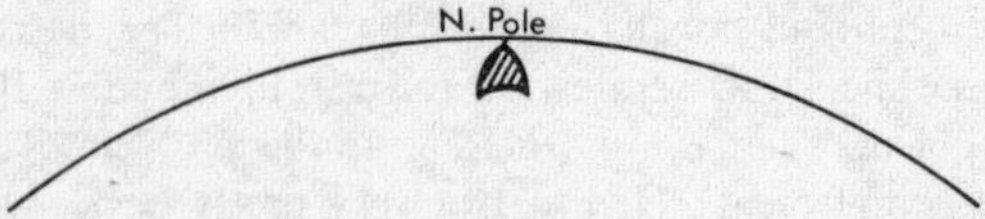

The bear hunt

In this problem our slogans are not helpful. There is no way of categorizing all problem solving. You need a flash of insight. Look at various positions on the globe. It is very satisfying when you get a solution. The North Pole. So the bear is white! Look at the diagram and we see that the hunter travels in a sort of triangle. The diagram exaggerates the size of the triangle, but the point is clear. From the North Pole there is only one way you can go – south. Once you have moved a distance – say a mile – then there is no problem about going east or west. Travelling along this particular line of latitude (as along any other) keeps us a constant distance from the North Pole. So if we came down a mile, travelled round a mile, then a mile north still gets us back where we started. The emotional reaction to solving a problem of this sort is a deeply satisfying amusement. It is funny; but it is nevertheless a challenging mathematical problem, and to solve it is comfortable, warming, a very positive experience.

Mathematics is deep and full of traps. There is a catch in this problem. For if you have discovered the North Pole as a solution the next question is 'Where else might the hunter have been?' This is a real 'sucker punch'. We have been drawn into believing that we have sorted it out and we are presented with the possibility of another answer (or perhaps more answers!). The resolution of this we leave to the Appendix (p. 250).

Attempts to measure intelligence are doomed to failure. Some will find number problems easy and those involving visualization difficult; others will find the reverse. We mentioned earlier speculations as to the capacity of the two hemispheres of our brain. One hemisphere is believed to control linguistic and numerical abilities; the other spatial perceptions. In any individual the imbalance may be very marked. The Soviet psychologist Krutetskii, discovered that it was present in school pupils of all abilities. Hence those problems we have posed on number and space will be seen as of different relative difficulty by different people.

The 'North Pole' problem is hard, not so much in finding the North Pole as one solution, but in discovering all (!) the other solutions – and there are many. Yet mathematicians (even if they could not do it) might claim that it was not important mathematically. In contrast, let us remember the 'Four Colour Problem' discussed in our last chapter. This problem, unlike some others, turned out to be mathematically significant. It led into a whole series of problems in an area called 'topology'. It is not uncommon for one problem to raise others and if these seem to form a connected area – an unweeded and uncultivated patch of knowledge – it suggests that there is hidden mathematical treasure to be found. It is emotionally unsatisfying (though often unavoidable) to prove something, as happened here, by analysing a large number of cases. But we cannot guarantee that anyone will ever come up with something more aesthetically pleasing. Yet that is one of the mathematician's criteria for the solution of problems – that they should be pleasing and have a rightness about them.

An illustration will help here. You are asked to draw a square (as large as

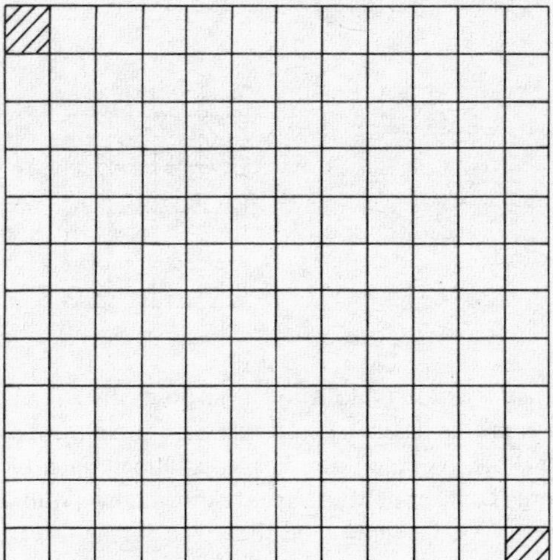

you like) and divide it into small squares, with an even number each way. In our diagram we have drawn 12 × 12, but we must remember that we have to deal with 'every' such square; that word 'every' may cause us trouble, as it did in some of the problems about prime numbers.

We are then supplied with exactly the right number of domino shapes to cover the area less the two diagonally opposite corner squares we have shaded in. In this case there are 144 less 2 or 142 unshaded squares. So we are given 71 dominoes. The question is, can we exactly cover the area using these dominoes? The fact that we have the right area to do it does not mean that we can necessarily make them fit.

In passing, it is worth noticing that although this is clearly a puzzle type activity, there are practical problems in the real world to do with economy in packaging and storing (imagine biscuits in a tin) which the thinking we are engaged in here might help.

Resort to our slogans. Always spend a little time clearing up any ambiguities in the question. We may ask our questioner whether the dominoes have to lie along the lines of the square. He will say 'yes' – but probably if he did not, we would come to the conclusion that we could not put them otherwise.

Suppose we have become quite sure what the problem is and we have already followed one piece of advice: we have drawn a diagram. How about 'TRY AN EASY CASE'? The lower diagram is 4 × 4 – with the corners missing. This is a good size to make a start. Mathematics is not, nor should it be, entirely an activity of the mind. Some may feel it demeaning actually to experiment, but that is a false restriction. Whenever possible, engage in a practical activity. Cut out the dominoes and the shape and try to fit 7 dominoes on to their shape so as exactly to cover it.

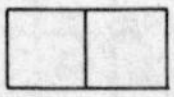

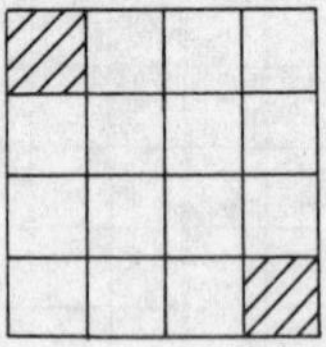

This problem was introduced for a purpose other than showing again how we go about problem solving, and enters into an almost mystic realm. As with the North Pole, its resolution lies in a sudden flash of insight, which however may only come to us after we have thoroughly familiarized ourselves with the problem. At the same time there is a further and very important slogan, not yet mentioned.

Rule 9: 'DOES IT REMIND YOU OF ANYTHING?

Some people seem able to rush to solutions of problems that we find impenetrable, and then to demonstrate to us how simple they are in fact. Those who are 'too clever by half' do tend to leave the rest of us somewhat diminished. Yet it does not follow, because they are swift, that they are necessarily deep thinkers. It may be this type of problem at which they excel, and the reason for the speed lies behind the slogan. Sudden insights that yield aesthetically economical solutions result from a wide experience of problems and the drawing of analogies between them. There is what Professor Richard Skemp, in *Intelligence, Learning and Action*, calls 'resonances'. As with a piano, a note struck elsewhere in the room sets a string vibrating in sympathy. so it is with problems that need divergent rather than convergent thinking. At times, perhaps in the more unusual answers, we arrive by what Edward de Bono calls 'lateral thinking'.

Now for the solution. The squares remind us of the chessboard. So colour them black and white. The two missing squares are of the same colour. A domino always covers one of each. So it cannot be done. Ever!

The brevity of statement is deliberate. Go back and read it several times. Reflect on it; take your time. Then analyse the feeling you experience. The economy of thought has genuine beauty. It may leave one feeling foolish, but do not think that because once grasped it is 'obvious' that many could in fact arrive at this insight.

So exciting can these insights be that it is worth offering another problem, this time without the solution.

You are given a block of wood in the form of a three inch cube. We want to cut it into its separate inch cubes by sawcuts. Again, there are issues in industrial processes that are not exactly this – but they have 'resonances'. If we hold the cube, as shown in the upper diagram on p. 198, steady in position, and do two saw cuts vertically one way, two another way vertically, and finally two horizontal cuts, we shall be reduced to our 27 one inch cubes. Not everyone finds this particularly easy to see. It depends on our powers of visualization, which vary greatly from person to person. The problem which we now state will prove quite hard for most people, yet there is a clue in that there is a strikingly simple way of seeing the answer. The unsimple thing is to find that way.

'If after the first cut we are allowed to rearrange the pieces of the cube (as for instance in the lower diagram) before we make the second cut, and then to rearrange again between cuts, can we do it in less than six cuts?'

There is an interesting language point here. Most people, when asked if they 'can' do it in less than six cuts, draw the implication that it is possible. The mathematician uses 'can', very properly linguistically, as an open-ended question. So here the problem is to show how to do it if it can be done in less

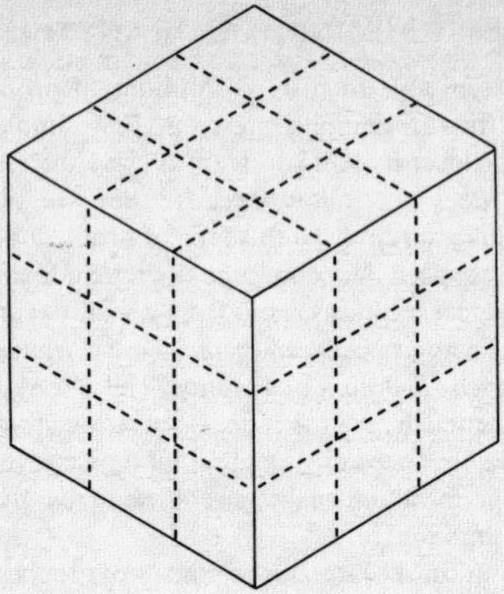

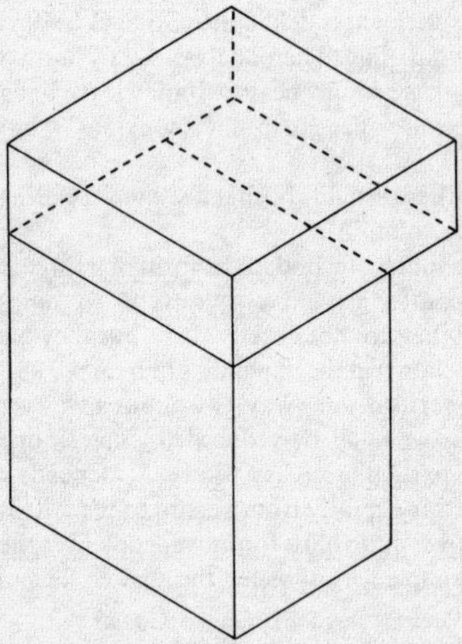

than six cuts, but also to demonstrate why not if it happens to be impossible (as in the chess board problem). A solution is given in the Appendix (p. 250).

Problems then have various parameters. They may be easy or hard, important within mathematics or not, and with practical applications or not. Some people claim that it is only those which are 'useful' which will motivate people to attempt them. Experience does not support that view. Certainly it can enliven and enrich the study of mathematics to realize that it can be directly applied to 'life' problems, but the attraction of puzzles of a mathematical nature offered in many mass circulation journals suggests that there are deeper psychological factors at work which determine whether we engage with a problem or not. This would be an interesting hare to pursue, if only enough were known about it to offer a coherent range of thoughts.

The move from algorithms to problems leads to a great opening up of both mathematical appreciation and of our own internal thinking processes. In educational terms, it is this interaction between thought and subject content that is so valuable. The further step to 'investigations' is another fascinating one. It can leave one feeling even less secure than before about mathematics, but once you overcome this fear, and work by yourself with no one looking over your shoulder, new and amusing worlds may open up.

Let us start an investigation. It does not finish. That is one of the main differences between problems and investigations. Generally speaking, when you have solved a problem, you know it. Draw a line across a sheet of paper (Diagram 1). Since we are engaged in mathematics, we shall assume this line and the paper it is on both go on for ever. If the notion of infinity is a trifle disturbing, then perhaps we could modify this to going on 'as far as we like'. We shall see why we need this proviso in a minute.

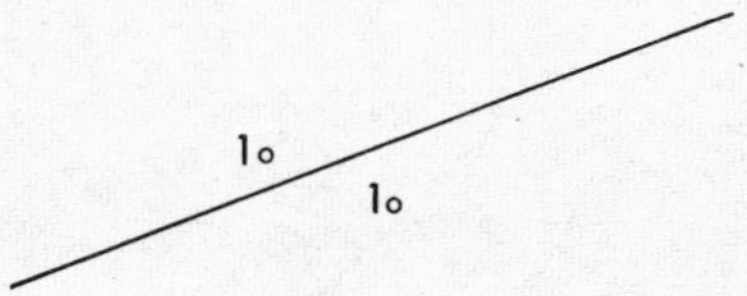

(1)

The line divides the plane of the paper into two open regions. The nice thing about some investigations is that they start from something which is simple and obvious. We might not have thought it worth commenting upon. The two regions have each been labelled 1o. That may seem slightly mysterious until you are told what it means. The words on this page are mysterious to someone who does not read English.

Mathematical symbolism and the troubles it causes deserve another book. Yet here we are in control. Symbols shall mean what we choose. '1o' here means it is a region with one boundary and it is open. That is all.

Diagram 2 has introduced a second line (not in any special position). Provided it is not in a special position, such as being parallel to the first line, it will cut it in one point. Here we can see why we made the proviso about extending them as far as we like. If we drew the lines at random they might run off the page before they met – but they still meet.

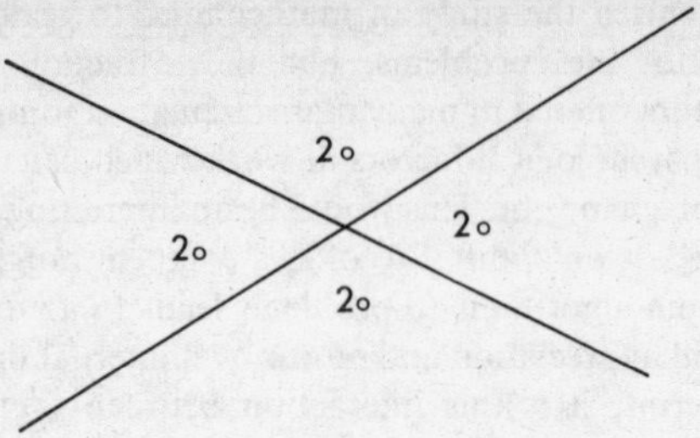

(2)

Using the 'notation' we have devised, we how have four regions, all open, all with two boundaries. Open, of course, simply means that they run off for ever. We have labelled each 2o.

Move on to three lines and we are in Diagram 3. We still have a niggling thought about what might happen if the lines (or perhaps only two of them) were parallel. What if they all ran through one point? Or if one were on top of another? These are 'special' cases and lead to a question we shall flag for now and look at later.

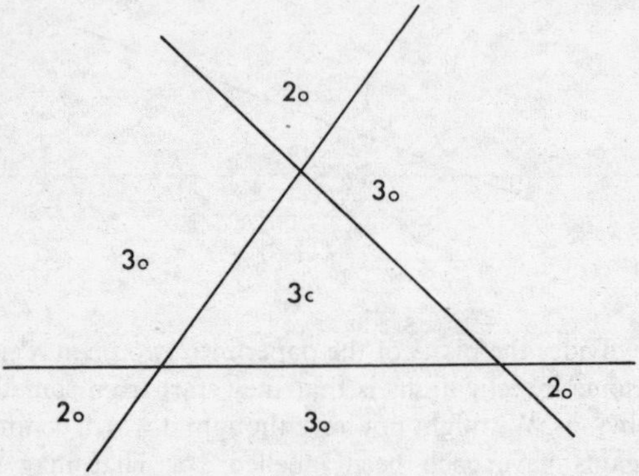

(3)

'How many special cases of three lines are there?' (i) (Answers to some of the questions posed will be found in the Appendix, p. 251.)

In Diagram 3 we are looking at the general case, where all the crossing places are distinct and separate. The regions have become just a little more complicated. We have three of the 2o sort, three more of the 3o and one labelled 3c. the last one is a closed region with three boundaries. Just a triangle, you may well think. Indeed it is. Yet it is a triangle placed in context.

In mathematics we usually think of lines as going on in both directions for ever. So if we had started by drawing a triangle, we would have ended up with Diagram 3. And why should we only pay attention to the inside of the three lines, and not the other regions we get?

There is no prize for guessing where we go from here. We put in a fourth line, crossing all the other three. We have a passing thought which we may record

'How many special cases are there with four lines? (ii)

But then we move on to look at our new configuration (Diagram 4). It has been labelled in the notation we have devised.

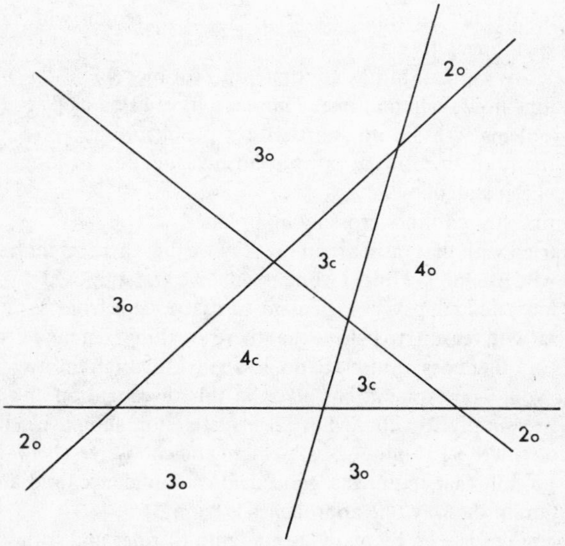

(4)

We start to wonder. Is that the only configuration of four lines we could have drawn (leaving aside the special cases, which we have already decided to look at)? Rephrase that in slightly pompous mathematical terms.

'Is the 4-line configuration unique?' (iii)

At this stage we might draw up a table of our observations thus far. We will list the number of lines against the types of region and indicate how many of each sort we get.

We are now well into our investigation. As we go on we realize there are all sorts of things we do not know. Can we guess at the answers that will appear in our table for 5 or more lines?

Lines	1o	2o	3o	3c	4o	4c
1	2					
2		4				
3		3	3	1		
4		3	4	2	1	1

Stop here and ask some more questions.

'Can we predict anything about the types of region with any given number of lines?' (iv)

'How many configurations are there of 5 (or more) lines?' (v)

Questions flow thick and fast. Those we have listed as (i) to (v) are all genuine problems. We did not start with a problem, but we seem to have generated many of them. That is what an investigation is about. This one could have been stated:

'Examine straight lines crossing in a plane.'

Had we started with that rather than the way we did, there might have been a sense of bewilderment. We might wonder what we are 'supposed' to do. This is one of the most difficult psychological postures to escape from. We are mostly conditioned (with reason) to believe that there are things demanded of us, that the teacher or other boss wants us to do. In an investigation the way you go is a matter for you. The explanations given in this investigation and the paths suggested are simply to show what an investigation should be. Ideally we should start with the 'Examine . . .' statement and choose, freely for ourselves, where we go, but that requires a good deal of confidence, and a degree of sophistication in the way one approaches learning.

Mathematics is seen by many as made up of rules and formulae which constrict and constrain. This is because their main experience has been with algorithmic processes. Mathematics, properly understood, is an area where questions and issues needing answers constantly arise. Therein lies its magic. A further feature of an investigation is that it can provide interest for both the mathematician and for the tyro. Any reasonably intelligent person will find out something in this investigation, yet there are issues in it that may well prove beyond the most able mathematician.

At the easier end lie questions such as

'Can we predict the number of open regions with any given number of lines?'

'How many crossing places are there?'

Applying some of our problem solving techniques, it should be possible to find an answer to these.

However, as a challenge to a mathematician, how about this:

'Derive a formula for the number of different configurations in an n-line diagram'?

Here we are into a piece of mathematics that has a beginning but no apparent end, raises questions at many different levels of difficulty, and is based on some very simple starting points. We may also wonder if it has any practical relevance. It may not matter to some of us whether it has or not; others may have reservations about ivory towers.

The relevance may not be immediate, but the style of thinking it encourages may be valuable in 'real' situations. The networks of lines we have been working with have been purely geometric, but many human artefacts are networks of lines. Consider the streets in a town, an underground train system, or an electrical circuit. All are basically the same as one another and all are like the system of lines we have been looking at. Some of the problems in the investigation may well have corresponding problems in traffic flow, train timetabling or wiring a house. A feature of mathematics is that its very abstractness lends itself to applications in so wide a range of human activities.

Another investigation leads us into a new and developing mathematical field – graph theory. This has been discussed in our previous chapter, but one problem from it leads into another multifaceted investigation.

Take three points at random in a plane. Join them by straight lines that stop at the points, and do not extend indefinitely as before. Generally the lines will be of three different lengths. It would be a special case if they were not. Yet if we were free to arrange the points as we chose we could achieve a picture where there was only one length of line. In the second triangle there are still, of course, three lines, but they now all share a common length, forming an equilateral triangle.

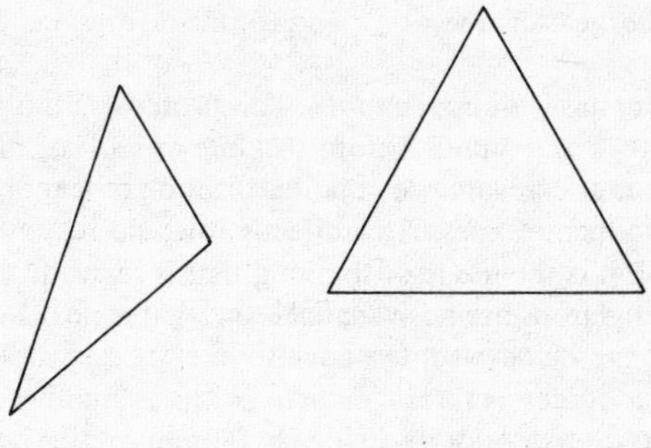

Once again we have commented upon the obvious. As we progress the issues become rapidly less obvious. With four points we have six lines, all possibly of different lengths. Noting as we pass that we ought to be able to forecast the number of lines, given the number of points, we try to reduce the six different lengths to rather less.

The first inclination is to seize upon a square. Now the six lines give only two different distances. One is the length of the side of the square, the other the diagonal length. At first sight we have been faced with a moderately simple problem, and solved it. Beware of settling for a single answer. Again our conditioning has been at work. The sort of mathematical work we have previously encountered has led us to believe that there is generally one single answer. In this case there are two other arrangements which also mean that there are only two different lengths involved. So our next question is to find them. When we have, can we be sure that this is all there are; how can we establish such a fact?

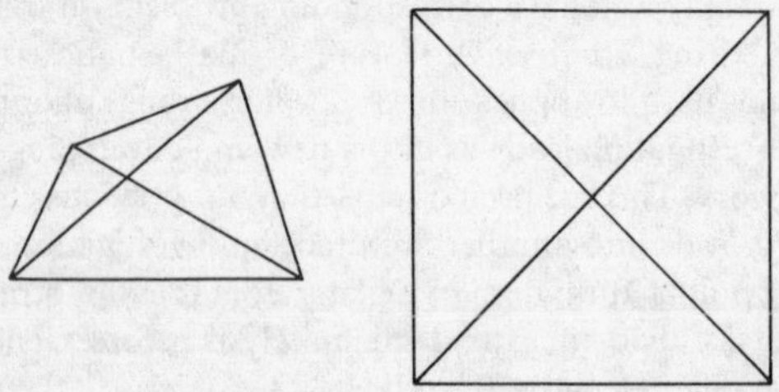

As before, we have only set out, and are faced with stumbling blocks all the way. The issue about whether there is a general rule for arranging the dots to minimize the number of different lengths needed is an unsolved problem in graph theory. Yet we are free to investigate different systems of dots, perhaps set ourselves other, easier questions, and to pursue any line of thought that occurs to us.

Graph theory has wide applications. This problem is entirely a useful one in which to exercise our minds before tackling any of a range of practical problems. There is one 'resonance'. Standardization of length may well be an issue in some production lines. How directly it would relate to this problem may be doubtful; it is the mode of thinking that is relevant.

These investigations are both spatial – but that is not of course the only area where they can be devised. Go back to one of our problems in number theory – the Syracuse conjecture. We started with any number and, depending on whether it was even or odd, we divided by 2 or multiplied by 3 and added 1. There is not a lot of reason apparent in that. It just happened that those two rules seemed always to bring the string back to 1. Here again, we must free ourselves from our instinctive acceptance of being rule-bound. It is we who make the rules. This time we might say, 'Examine strings of numbers formed by certain rules'. There used to be a fair degree of tedium in working through such investigations. With a microcomputer at one's elbow the donkey work disappears and the amusement remains. A short program on such a machine will allow you to check the Syracuse conjecture for the first, say, thousand numbers. The machine may take an hour or so, but it does not get tired, and if properly addressed, will tell you which numbers produced the longest chains. What the human mind does is think of things to look at. The starting point is something like 'I wonder . . .' or 'What if . . .'. This speculative, questioning, creative state of mind is characteristic of genuine practitioners in both science and mathematics. It is not the general view held by others of what these studies are about.

Taking one of our own pieces of advice, let us reflect on what has been said in this chapter. Mathematics contains many algorithms, it is a rich source of problems, and a region in which we may conduct many investigations. The distinction between them has little to do with difficulty. An algorithm for solving differential equations may be complicated and may need much prior knowledge within the subject to use it. Ultimately, however, it is routine. Algorithms will eventually be entirely the province of machines. A problem may be trivially easy and require little previous mathematics to do it. What makes it a problem is that a method had to be devised for its solution. None was automatically to hand. Of course, what is a problem to one person may be routine to another. If there were a person who had never heard of any method of long division, and they were faced with the dividing of one number by another, that to them would be a problem. Within mathematics there are many known routines for many processes: all were once problems. They have now moved into a 'dead' area. That is not to say thay are not used – they are – but they are static.

We have seen that problems easily stated may be impossible of solution, or perhaps we should say have not been solved after centuries of effort. Had their impossibility been proved, that in itself would be a solution. That may seem a paradox, but a famous problem illustrates this point.

The Greeks found many geometrical constructions, including, as a very simple one, a method for bisecting an angle. Many of us will have learnt this at school. A restriction we now see as artificial – that we may only use a straight edge and a pair of compasses – was imposed on all these constructions.

It was perfectly natural to ask the next question; can we trisect an angle, or cut it into three equal parts? The movement from the solution of one problem to the formulation of another is characteristic of an investigatory approach. Many attempts have been made on this task. 'Solutions' are still presented by enthusiastic amateurs in the subject, and many reveal much ingenuity. Yet it is over a century ago that the impossibility of the construction (with the constraints stated) was established.

Oddly, it came not through geometry, but through algebra. Briefly, if the problem turned out to be a quadratic (that is involving x^2) then it could be done with straight edge and compass. If of higher order, it could not. Trisecting an angle is equivalent to solving a cubic equation (one involving x^3) and it cannot therefore be done in the terms stated. Mathematicians had therefore arrived at a 'resolution' of the problem, not a 'solution'. Yet it told us much and showed deep relationships between algebra and geometry.

Investigations and problems are intimately entwined. Problems once solved can open out into investigations, and an essential feature of an investigation is that it is problem producing. We can, however, define distinctions by way of a diagram.

With an algorithm we know where we start, we know the goal, and the algorithm is the method for moving from one to the other. It may not be easy, there may be snags in the route, but the tracks are laid down. A problem is characterized by knowing again where we start and where we need to get to, but having to build the track. In an investigation we decide where to start and we start building our tracks in any direction we like with a Micawber-like confidence that something interesting will turn up.

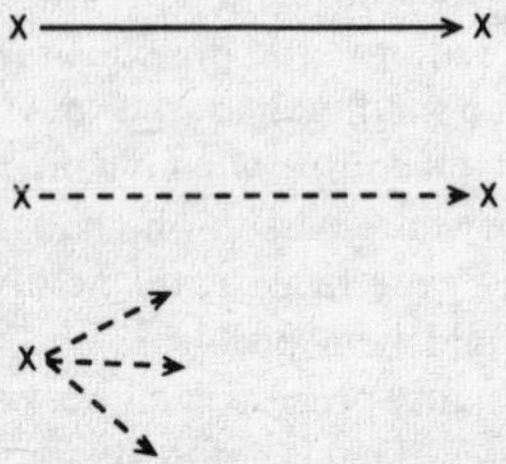

Progression of algorithm (top), problem (middle) and investigation (bottom)

We need to learn algorithms – or at least some of them – but to experience what mathematics is about we must solve problems and, hopefully, conduct our own investigations into the subject.

IV

Summing up

13

Mathematics: its nature and purpose

Society has always decreed that those who undertake education should have a substantial input of mathematics. In Britain there has been compulsory free education for all for well over a century, so that nearly everyone now living has had at least nine to ten years of mathematics, commonly every school day they attended. This is a large investment, and may well need justifying. That is not to agree with those who claim that things are much worse than in their day; the evidence is all against them and needs no discussion here.

Among the pupils there is a curious ambivalence towards mathematics. It is generally recognized as useful and important, and we shall examine these claims. An early report by the Schools Council showed that among 15-year-old school leavers, who on the whole would not be those who most enjoyed education, over 90% of both boys and girls believed it to be useful; far smaller percentages liked it. Teachers accept its importance, yet many of those who teach the subject would struggle to give a convincing rationale for what they teach. When the pupils (despite their apparent beliefs) ask 'What use is this?' the answers provided are not always compelling. Yet our educational institutions state, always implicitly, and often explicitly, that the subjects that really matter are mathematics and our English language. They are the subjects that can be assured of ample time allowed them. Shortages of mathematics teachers has sometimes led to its losing ground on the timetable, but this is very seldom a school policy; it has been forced upon them.

Employers share this view. A candidate for a job with O levels in art and religious education is not likely to compete with another, still with two O levels, but in mathematics and English. Institutes of higher education look to these two subjects, no matter what course the student proposes. Many educationalists regret and seek to combat this view, but the weight of opinion is strong. To say that society holds strong views does not mean that they are right. Certain very regrettable views have at times been held by very large proportions of certain supposedly civilized countries. Because mathematics is widely regarded as 'important' does not mean we should not examine this claim in some detail, drawing on many of the issues we have explored in this book.

Mathematics for survival

If a person lacks certain capabilities, it can be crippling to him, both in the practical difficulties entailed and in the humiliation and embarrassment he may suffer. In our modern world, not being able to read is such a disability. One hundred and fifty years ago it would have been a matter of no comment. In saying whether a person can or cannot read we may need to define levels, but there may be as many as two million illiterates in our adult population. Those who deplore this and blame the system do not realize exactly how difficult a skill it is, nor the manifold reasons that may prevent a person reading. It is likely the problem will never be fully solved, despite the many advances in techniques of learning.

In mathematics there is not a single central skill similar to that of reading. Though many might see the multiplication tables in this role, we hope to have modified this view in our chapter on 'The Basics'. To decide what is essential rather than quite useful becomes a difficult exercise, and where we draw the boundaries may readily be argued about. A useful criterion for defining a central core that is necessary is that of 'personal autonomy'. If, as we go about our everyday life, we need constantly to refer to others to help us through the day we lack this personal autonomy.

An inability to read may put a person in this category, though there is a telling riposte to this in a story by Somerset Maugham about a churchwarden sacked by a new vicar because he cannot read. He had some savings, invested them in a shop, gradually built a chain of shops and became wealthy. On learning that he cannot read, the bank manager asks where he might have reached had he been able to. The man replies that he would be earning a pittance as a churchwarden. Perhaps the moral of this story is that while he was not literate, he may well have been numerate. Human beings are wonderfully flexible in the face of many sorts of disability, yet there are some mathematical needs that it is very difficult to do without. They are fewer than we might believe.

We need to tell the time. Societies did not always demand this; once it was enough to say one would be at a certain place 'in the middle of the day' or 'when the sun goes down'. We are heavily burdened with time in Western civilization; it is the source of much of our stress. Yet if we live in that world, we have to learn to tell the time. Most of us did so on the old-fashioned clock-face, with rotating hands; these are now retreating under the onslaught of electronic digitals. We learn so early in life, that we do not normally analyse the skills needed. Two, sometimes three, arms rotate against the background of a circular dial. Time is measured by angular displacement, yet the circular dial we use needs different scales for the hands, which rotate at different speeds. We tend to think of things we learn early in life as easy, and those learnt later as harder. It is not always so. Telling the time is hard, yet nearly everyone manages it before they leave school. Perhaps before long we shall all use digital readings. It will certainly be easier, but among other things we shall lose an important early experience of rotation.

We certainly need to be able to count, and to be able to read numbers. If it is a 36 bus which takes you home you need to be able to recognize the digits which form the numeral. We need to understand ordering to some extent. If shoes or dresses are sold with numbers to represent their sizes, it matters that you know a 12 is larger than a 10 (but that a 10 dress has no connection with a 10 shoe).

Calculation plays a part. In our chapter on calculation the suggested limits for mental work cover what we need for survival. Most of our calculation takes place with money. It is probably necessary, and certainly very useful, that we can check our change from a £1 or £5 note. These are specific types of subtraction sums that certainly are more easily handled, as all shopkeepers know, by counting on. Even this skill may disappear with the extended use of tills that record not only the amount paid, but the change to be given.

A degree of estimation, not only in money, but in weights and measures, is pretty important. An estimate of how much material you need for a job, how much food you will need for a meal, and so on, is something we all need to be confident about.

Many of our daily routines involve sorting processes. There are more inadequate families who fail in organizing rather than from sheer lack of money. We have discussed this need in our chapter on 'Modern Mathematics' and the importance of sets but we have not come to terms with ways of teaching people to sort, order, process and organize. It is perhaps because it is seen as 'common sense' and assumed that you either have it or do not.

It is difficult to analyse the basic skills needed in geometrical or spatial work. Nowadays we handle many mechanized devices, as for instance a large range of different locks. People have found themselves in awkward positions through being unable to open a lock, which can only involve a turn and a straight motion, however combined. If we go far in this direction we realize how disabling it can be not to have any idea of circuitry or of some of the functions of a car's engine. Some at least of this is pure geometry. Whether things are joined up or not is basic to topology and to the working of a car. Seeing what shape matches what is another such basic skill.

We all of us travel so much that the reading of maps and diagrams is rapidly becoming an essential part of everyone's equipment. Travelling by car we may need to use a map that actually represents the area to scale. On the London Underground or for Inter-city trains we use diagrams which convey the necessary information without being to scale. It is not a taxing geometric skill to find the point where two lines cross, but it is a necessary one if we are to get around on the tube.

We might find others; the aim here is not to provide an exhaustive list to suggest some matters and to provide criteria for judging whether a piece of mathematics is necessary or not. If a task is met often (such as looking something up on a table of numbers) then we need to do it ourselves and not ask someone else. There are other needs that occur less often and where it is appropriate to go to a specialist.

In seeking to define something, in this case 'mathematics for survival', it helps to say what is not included (in mathematical terms when we are defining a set, it helps to say something about its complement). Among the things not included, then, are operations with fractions beyond a certain level and pencil and paper work on the four rules beyond a pretty low level.

We have seen that we need not go beyond halves, quarters and tenths. Equally, nowadays there is simply no virtue in being able to multiply two three figure numbers. Use a calculator, having estimated the answer.

What is necessary for survival, and what is believed to be necessary, do not match up – but at least they overlap, so let us draw a set diagram to show where we are.

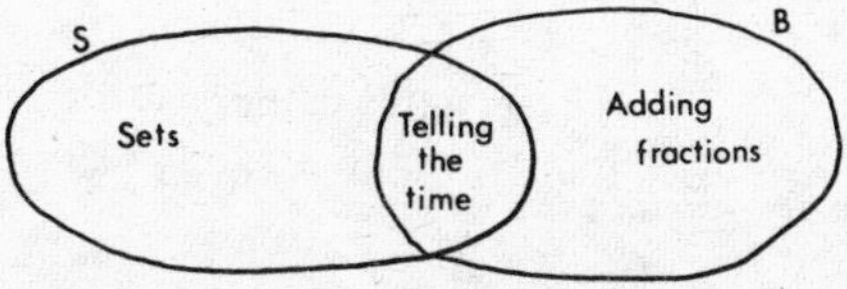

Survival mathematics: S = survival kit; B = kit believed necessary

'Useful' mathematics

Boundaries will inevitably prove fuzzy, and moving from necessary to useful is a very fuzzy boundary. Within the useful we would include a wide range of citizen's skills, certainly beyond the basics, but of value to the intelligent adult who wants to make sensible decisions.

An understanding of certain statistics, in particular the way people try to use them for their own ends, is of growing importance, More and more information comes to us through the mass media in numerical form, or in graphs which imply numerical facts, or even, as we have seen, in the use within language of such phrases as 'rising less fast than previously'. It is proper within an open society that people try to sell their wares or their political policies. We cannot always expect them to be totally honest in the way they present their case, so we need to have the means to make our own judgements.

It is useful to understand matters such as income tax. That is not to say we should all be accountants, but a nodding acquaintance with the way a banded system of taxing works should be in most people's armoury. We need to read and understand a range of dials, and perhaps to check them against our bills.

The degree to which we pursue this is partly individualistic. Different people have different attitudes to the degree of control they seek over their environment. Our chapter on the practicalities of maths takes us to the limits of what we might think 'useful', and sometimes a bit beyond.

If we wish to relate 'useful' to 'survival' we do so in our next diagram. All this Venn diagram expresses is that we would include our survival kit in what is useful but that some useful things go beyond survival. Remember that in these diagrams there is no implication to be drawn from the size of enclosures; the important thing here is that one lies within the other.

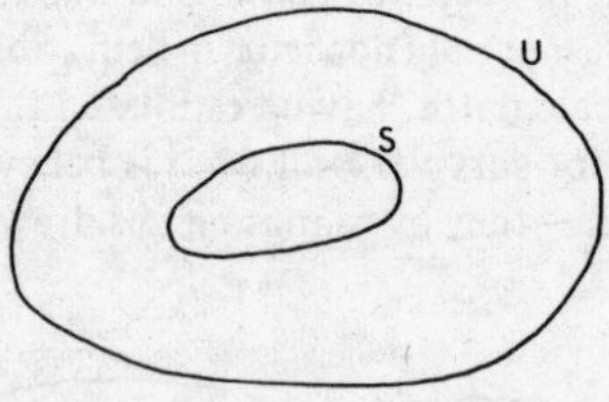

Useful mathematics

Direct applications of mathematics

In a sense the categories we have discussed so far are applications of mathematics to the world around us. This extended category we are now looking at is more technical. It is not material that is at everyone's finger tips. The main areas are those we discussed in 'Mathematics in Action' and the criterion we shall use for this admittedly arbitrary classification is that the mathematics deals directly with the problem, it is not called in as a tool to help another discipline. This was the way Galileo and Newton saw themselves working. They used mathematics to analyse motion, and in so doing Newton created the calculus. This branch of study became known as applied mathematics and was and is taught in schools as a part of mathematics. Some purists might feel it should more properly be part of physics. Naturally mathematics is used in it, but we shall shortly examine that role of mathematics as a separate issue.

More directly mathematical was the development of operational research, of topics such as linear programming and systems analysis, all matters involving process and how things are done. Here the situation needed analysing and the mathematics developed to meet the needs. Much the same is true of statistics, now regarded as a subject in its own right, but developing in response to a growing awareness and need to handle numerical information in an increasingly technological world. We have seen how topology and graph theory had roots in the problem of the bridges of Königsberg and how they now find direct application to the problems of society.

This is an interesting and distinct area, often with a very modern flavour to it. We cannot tell when the examination of some new question will start another proliferating branch of mathematics. Mathematics constantly grows and some of the growth is in response to the needs mentioned here.

We make a step forward with our Venn diagrams and draw yet another.

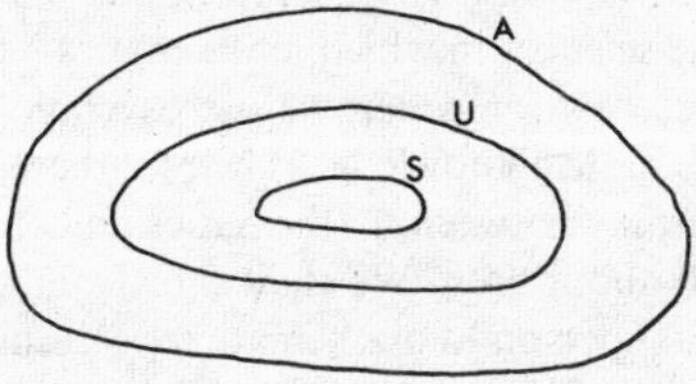

Applicable mathematics

Mathematics as a language and a tool

A physicist nowadays cannot pursue his or her studies without extensive mathematical equipment. The concepts needed to understand relationships in both the hard physical world of concrete objects, and the web of forces and fields that seem to predict and explain what happens, are all expressed in terms of mathematical symbols and equations. So far has the mathematics of physics gone that there is an uneasy feeling that the mathematical symbols *are* the entities which physics discusses. So complex are many concepts in modern physics that enormous and expensive equipment is devised to discover a trace of a rare particle whose existence would suit the mathematical equations in which it first appeared. This total dependence of physics upon mathematics as the language in which it is written had led people to believe that there is the closest philosophical connection between the disciplines; this is a view we shall later dispute.

The other sciences as they develop tend increasingly to depend on explanations written in mathematical terms. At one time it was permissible for a chemist to know little mathematics, provided he stayed away from certain parts of physical chemistry. That is rapidly becoming less possible. Certain areas bordering chemistry and physics have become highly mathematical. The work on DNA, brilliantly described in *The Double Helix*, can be viewed as inspired geometry. Biology has long since moved from its descriptive phase and parts, such as genetics, have strongly mathematical notions embedded in them. Geography, perhaps because of its increasing use of mathematical treatments, is more and more regarded as scientific, a view less prevalent fifty years ago.

We have been treating this as if it were a modern trend, but here modern thinking is echoing the thirteenth century. For instance, Roger Bacon said: 'Mathematics is the gate and key of the sciences. Neglect of mathematics works injury to all knowledge, since he who is ignorant of it cannot view the other sciences or the things of the world. And what is worse, men who are thus ignorant are unable to perceive their own ignorance and so do not seek a remedy.'

It is not only the sciences in which mathematics has become important. The philosophical reasons for which we list psychology separately will be given later but mathematical knowledge is helpful there not only in statistical work but in the interplay between mathematics and the cognitive processes that teaches us much about both. We shall shortly enlarge upon this.

Boole was the first person to mathematicize certain aspects of language, and with marked success. We have looked at this in 'Modern Mathematics'. Much more recently the work of Chomsky and others has lent a markedly mathematical flavour to the study of linguistics. The form and structure underlying any language and the permissible transformations within it are a natural field for a mathematician to enter.

Naturally it was mainly mathematicians involved in the first thrusts in computer science which, like statistics, is now a flourishing discipline in its own right (which threatens to engulf everyone else). Again there is an intimate relationship with language and with the patterns of the mind. It may be that this move, of greater potential than many developments in the intellectual history of mankind, will re-emphasize that the central nexus is between mind, language and mathematics.

All the disciplines of which we have spoken have their own regions of study, their own range of problems. To those who practise them mathematics

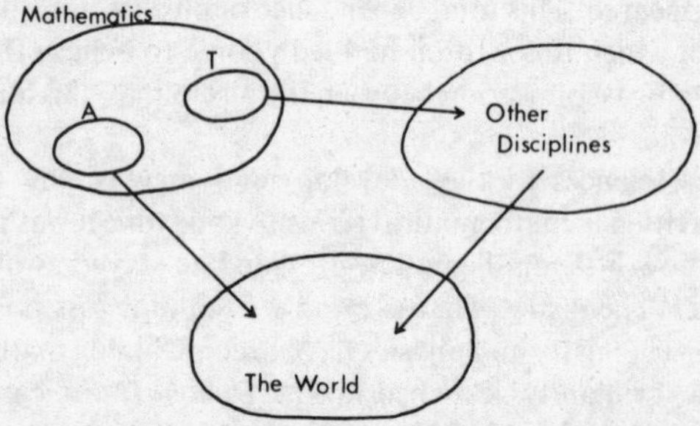

Mathematics applied to the world (A) and used as a tool in other disciplines (T)

may well be seen as a tool or a means of communication. To them that is what it is. In our analysis this is simply *one* of the things mathematics is about. Our next Venn diagram shows the relationship of applicable maths, the mathematics used as a tool and the real world.

Mathematics for its own sake

Mathematics is a complex organic growth with many internal relationships. While we think of applicable mathematics as a change in the structure brought about by outside influence, we must also consider changes wrought from the inside. It is possible to work entirely inside mathematics, oblivious of the outside world, and to make great discoveries within its boundaries. It is difficult to know if this is possible in other subjects. Most of them seem to be about other things, not about themselves. Mathematics is capable of being highly narcissistic. It may be that linguistics is similar in this respect. Once language has been created it is possible to work within it, not necessarily even being concerned with meaning.

We have fully established the great effect that external influences can have on the development of mathematics, yet the fact that it does not need this influence is a singular feature. Some creative mathematicians have been motivated by a desire to solve problems deriving from the real world; others have no interest in it – the mathematics itself sufficies. One has to have a highly abstract mind even to appreciate the sort of thing that can go on. A very pure mathematician would be content to start thus:

> 'Consider a set of elements a, b, c, . . . and two operations * and /. They conform to the following rules. . . . '

He then goes on to say what is and is not permitted. If you ask what these elements a, b, c, . . . are, he will reply, 'they are defined by the connections enshrined in the rules'. If you say, 'What exactly are the operations?' he will reply, 'The rules indicate what the operations are in terms of the elements.' In other words elements and operations define each other by the rules governing them. The system is totally self-referent. It seems we may be in a 'black hole'.

That is not quite true. We have already given examples (the Greeks with conic sections and Grosseman's algebras) which were developed as a pure mathematical exercise and later found uses in the real world, thereby escaping from within mathematics. Perhaps the main point is that their creation had no external impulse, nor did it matter if they related to the real world.

We have drawn our latest picture (p. 216) with the moles burrowing inwards, but the links with the outside world may be more complex than our diagram suggests.

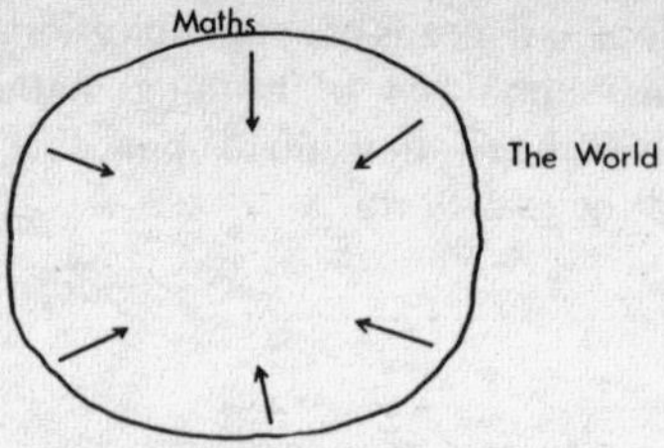

Mathematics for its own sake

Mathematics for personal development

Crudely expressed, 'Mathematics is good for you'. There is a degree of acceptance that mind training of one sort or another is offered by certain subjects. It was widely believed at one time that the study of the classics 'trained the mind'. This view was supported by the number of top brains with classics degrees. The cynical might observe that if those seen to be the most able intellectually were funnelled into that particular study, then the resulting excellence of those with classics degrees does not imply that the study of classics caused the excellence.

At any rate, the study of structure apparent in these languages may be less potent than the acquisition of new process. Thinking is a process, not an appreciation of structure. Modern linguistic approaches to language do, of course, see it as a process, as well as a structure. Let us illustrate the matter of process by an example we have already studied. Euclid's proof that the number of primes is infinite started with the conditional: 'if' they stop, then what are the consequences? One consequence was that there was a higher one and we had a contradiction. Hence they went on for ever.

Knowing this fact about the primes does not help one think at all. Yet if the necessary *process* had not previously been understood and is now available, then that is a marked advance in what our minds can do. They have a new form of reasoning open to them. Anyone who has worked in mathematics knows the deep joy of a new insight, a way of working and of seeing things new to one's mind.

The processes of mathematics and of logic are the processes of the mind. In learning mathematics we are matching our minds with the external manifestation of minds more powerful than our own. It is not exactly that mathematics trains the mind. Its processes are those of the mind. In the words of Kelvin: 'Do not imagine that mathematics is hard and crabbed and repulsive to common sense. It is merely the etherialization of common sense.'

In justifying mathematics, overkill is easy, since no one is resisting, but the reasons why we learn mathematics and what it can do are worth sorting in this way. In doing so we have expressed views about its nature, some of which will now be expanded.

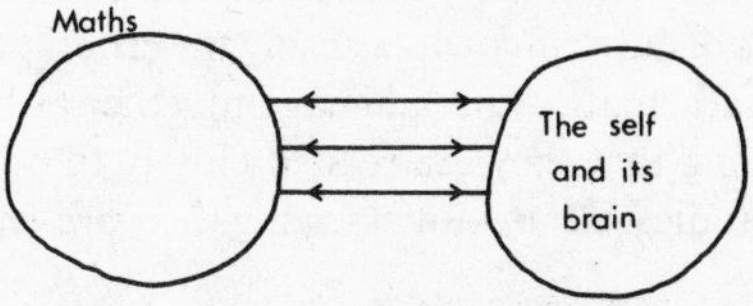

Mathematics for personal development

Mathematics and other disciplines

The bases for this discussion involve something of a diversion, so we shall summarize the main issues here and give a more extended account in the Appendix (p. 254ff.).

If we take two questions for each discipline and ask 'What does it discuss?' and 'What tests do we apply for its validity?' we arrive at some interesting classifications. Physics and history, for instance, both discuss the real world, though different elements of it, and the tests of their validity lie there. Mathematics and language are the medium in which these studies are written. Their tests are internal, and to some extent they are about themselves. In this sense they are closely allied, in another they are less so. An alien intelligence would need to communicate in mathematics; it would not share any other language.

The conclusion of this section then is that mathematics is of a rather different quality from other studies. It has on the one hand a universal standing (which language certainly does not) but it does not depend on that universe. It seems to be a rather curious animal.

Mind, mathematics and language

We need to explore rather more the deep connections we asserted between the way the mind works, and the structures of mathematics and language. The analysis in the Appendix concentrates on what in schools are classified as 'science' rather than 'arts' subjects. History, for instance, is clearly about something in the real world, not in its physical manifestations, but certainly in happenings. It could be said therefore to have deep connections with physics. The direction in which history increasingly may move is into contemporary events and then into predicting what may happen in the future. To conduct this study, we use the vast reservoir of written material about what has happened. It is only because man 'uttered' so many things (not only written) that history is possible. History is mainly written in language. That does not make it language any more than physics is mathematics. The study of language is, like the study of mathematics, self-referent. If, therefore, we are to pursue classifications of curriculum areas using the criteria we have stated, it is natural to classify mathematics and language together as being distinct from other studies. In saying that they are related closely to one another and are distinct from other studies, we are not asserting their pre-eminence, only their difference.

If we accept that they are not about external matters, and that at least to some degree they have internal references distinct from other subjects, we need to know now whether there is a special relationship to our ways of thinking. It is sensible to start early on with the human animal, and look at how a child acquires language and early concepts. Chomsky discovered deep structures in the various languages of the world that showed some very basic similarities. This fact, together with his belief that the language structures which young children acquire were so much more complex than their cognitive processes could be expected to encompass, led Chomsky to believe in an internal pattern in the mind, genetically transmitted, that enabled this remarkably fast acquisition of language. Criticisms have, of course, been levelled at this stance. Margaret Donaldson claims to demonstrate that children do in fact have reasoning powers so much greater than we believe, that the assumption of a genetic imprinting is not necessary. We need not pursue these arguments. We are concerned to show that the patterns of language are a close match for the patterns in our minds, whether or not they are genetic issues. If we express our thoughts in language, does this directly imply that thought and language are isomorphic?

Let us now recall Boole, whose attack upon language (to reduce it to algebra) was later interpreted spatially in the Venn diagrams which we have used extensively in this book to explain ideas. Perhaps 'explaining' means putting in a form where the mind can take it in and fit it comfortably. Since Boole was particularly concerned with logical processes we see him as unifying thinking, speaking and mathematical symbolism. His methods might be seen as translations from one language (English) to another (mathematics). Yet the subject he created, whether it be thought of as logic or, as Russell claimed, the beginnings of pure mathematics, seemed not to be of quite the same quality as mathematics, or perhaps we should say it is in a different universe of discourse. If we take the whole body of mathematics, logic seems to be about the way it works, and logic is the way the brain works. It may well be that linguistics forms the same sort of bridge between mind and language.

There are no proofs of position in the way we look at these sorts of relationships. We are content if people come to believe that mind, mathematics and language form an intimate central core to the things we humans are about, whatever the role of our other activities.

Mathematics, mind and emotion

Perhaps because of the connections we have asserted, mathematics can play a particularly important role in studies of how the mind works. More than in most areas, the reasoning processes are clear, defined and explicit. If we feed into a mind pieces of mathematics, we can study the manner in which it is processed. We illustrate this first with a distinction between instrumental and relational understanding, a notion initiated by Stieg Mellin-Olsen and developed by Richard Skemp. In testing what the mind knows, we are obliged to look at its output, and if correct answers are given it may not be easy to say what the mind knows.

Start with this statement:

$$23 \times 10 = ?$$

If we offer it to a variety of people, many may produce the answer 230. This does not mean that all have the same understanding. Offer a series of 'explanations' to those who get the correct answer, and see which is chosen.

(1) When you multiply by 10, you add a 0.
(2) In our place value system, the columns increase by multiples of 10 as we move to the left. Move them one and fill in a place holder in the units.
(3) Twenty-three lots of 10 is 230.
(4) Because 10 has a zero, the use of the zero will increase the number ten times.

Consider the explanations, and try to be sure which you think best matches the way you see things. We analyse this in an appendix.

The nine times table is another example where we detect levels of understanding. Again, all the people we test may get the correct answers but we can define levels none the less.

(1) Those whose replies are simply learnt responses.
(2) Those who have seen the pattern of the table. Look at the numbers 18, 27, 36, 45, 54. . . . We observe the digits always add up to 9. Also that the tens go up in ones and the digits down in ones. This is pattern spotting; and an important step towards understanding but not understanding itself.
(3) The understanding comes from the number line and the realization that a step of nine takes you into the next decade, but you fall back one. So the 10s increase by 1 and the others decrease by one – keeping the total at 9.
(4) A mathematician would offer a formal, algebraic proof of this – but let's not worry.

There are four levels of response, indicating the depth of understanding.

In the Appendix (p. 240) we describe an experiment with children, testing their knowledge of subsets, but we may have said enough to establish our starting assumption – that feeding pieces of mathematics into the mind and asking for structural responses can enlighten us as to the mode and level of understanding of that mind.

The actual content of mathematics is emotion-free and is not value-laden. Since the testing of mathematics is internal, we are not seeking after truth. (Even if it were not, we might not find it – as the man once said 'What is truth?') The issue is consistency. This point needs to be pursued before we return to the question of emotion.

A common feature of mathematics is axiomatic systems. The classical system is that of the Euclidean theorems. Starting from certain assumptions (Euclid called them 'self-evident truths'), a series of results were derived by logical process. The belief was that if the axioms were 'true' then the results were also true. Our present position is different. The criterion for axioms is that they be independent one of another. A consequence of this is that if we reverse one, the system must remain consistent, since it cannot affect the others. We thus not only do not have self-evident truths, but we have an assertion that either an axiom or its reverse will suit us equally well. It is this that led to Russell's oft-quoted remark and to the statement that the subject does not involve truth or value-judgements.

We are thus faced with cold rationality encapsulated and without the need for external reference. We now have what may seem a paradox. The subject least concerned with emotion creates among its students more anxiety and heavily negative feelings than any other. We cannot measure this, but surveys of adults give plenty of evidence. There are several reasons for this, fully explored in my *Do You Panic About Maths?* (1980) but, briefly, the failure of teachers not to consider the emotional dimension while teaching mathematics (on the grounds that it is emotion-free) together with the authority and time pressures that always seem to be present, reduce many people to a very unhappy state in the subject.

That is not to say that it is only negative feelings that are generated by the study of mathematics. Russell speaks of the 'delicious' feeling of first reading Euclid, akin to first love. While this is not exactly typical, most of us get genuine and very specific feelings of elation when we either solve a mathematical problem or when a new understanding of a piece of it slots into place within our minds.

Sex differences in mathematical achievement

We may here be straying slightly beyound the nature of mathematics, yet if it did happen that the patterns of men's minds and women's had some differences, then the way in which each reacted to the subject may say something about it.

The facts are that far fewer girls take the higher school examinations in mathematics, and that it seems that this is a general phenomenon around the world. In England, there are marked differences in achievement at O level and far fewer girls study mathematics in the sixth form and take A level.

An equally important fact is that in such testing as is done across large numbers of the school population at earlier ages, the mathematical performance of boys and girls is pretty equal, with girls perhaps having the slight edge. The evidence is confused by the fact that girls mature earlier and that intellectual spurts sometimes seem to accompany this. Even splitting the results down into topics seems to show no marked differences, despite studies in some countries that spatial abilities are lower in girls. These studies are not reliable, however. The general fact seems to be that there are not significant differences. Recent testing by the Assessment of Performance Unit in England detected differences in attitude in that boys of eleven were more confident than girls (even when they were wrong). This test was not conducted in other subject areas, so the result tells us little, for the sexual stereotyping, which we all accept goes on, demands expressions of confidence from boys in all matters (which does not necessarily mean they actually feel it). The author's researches into adults suggested that many men were quite as anxious as women, but were slower to admit it. In any case, let us concentrate on measured performance.

Thus far we have pointed facts about children and young adults across the whole range of ability. If we look at that very small class, the creative mathematicians, those who create new mathematics, the situation is very one-sided in the men's favour. That is another fact that our speculations need to explain.

The first question is the possibility of genetic differences between men and women. This occasions great heat in some quarters, even if one shows a willingness to consider it. Clearly it weakens conclusions if some possibilities are not explored, and are ruled out of discussion from the beginning, yet there are people who would wish to start from the assumption that there can be no difference.

The reason for the heat lies in questions of superiority and inferiority rather than the difference. In an ideal world, such questions should attach only to moral issues; the issue of whether one person is 'cleverer' than another in a particular area should matter no more than whether they are physically bigger. Realistically, intellectual ability in any sphere does lead to greater marketability of oneself and eventually more senior positions (at least, statistically), so it matters more to be brighter than to be heavier.

It is certainly not impossible, a priori, that men and women should have different built-in potential for doing mathematics. They are physically different, and may be mentally. The paucity of great women mathematicians could arise from such a cause, unless another is evident. Certainly another is evident. It is only in the most recent history that women have remotely been offered equality, and there remains a long way to go. With such heavy social conditioning it would be unlikely that they would figure largely in any major creative work – and this has been seen to be so. The fact that it is so also has the effect of establishing that there is nothing particular about mathematics, so the evidence of few women mathematicians does not relate to specific issues in mathematics.

This historical situation, that much of what has been created has been created by men, could lead to another possibility. The game of chess is played very largely by men, and very few women can compete at the highest level. A possible explanation here would be that if the minds were patterned differently (not with one necessarily superior), then the products of one of these types of mind would be more suited to that same type. This would be a convenient explanation for chess, but not for mathematics, which has a status in the external world. Despite its 'independence of experience' it could be said that the universe is constructed so that no one with sufficient intelligence could fail to develop it. This puts it in a different category from chess. The suggestion that men are better at it than women because men created it does not stand up.

The real evidence against basic genetic differences in mathematical ability must surely lie in the facts from which we started. If the evidence is that up to the age of thirteen there are no important differences in performance in mathematics, surely this must be solidly against genetic factors? In the face of this evidence, the onus is on those who believe that differences in performance in mathematics are genetic to prove their case.

In an issue generating much heat, bad arguments will be used on both sides. One 'explanation' of girls eventually performing worse, particularly in spatial matters (if that is proven), is that they are not given the same toys as are the boys when they are young children. There seem to be two major faults with this argument, which is constantly paraded. The first again lies in the facts. No one denies that heavy stereotyping is induced by the things used in early play, but if it affected them so strongly, why is this not evident earlier? The second is that playing with a doll, dressing it, or working with a doll's house is certainly spatial experiences, sex-stereotyped or not. Arguments directed at early experiences have then to explain why differences manifest themselves so late.

Much has been written in the United States about the negative effects of the elective system that operates in high schools. Social pressures result in girls doing much less mathematics than boys in their courses, and this is said to explain the differences. Pressure is then brought to encourage high schools to get the girls to opt for more mathematics. A mere glance at the English system would show that this is no solution. Effectively all boys and all girls take mathematics to the normal school leaving age without choice. We have the system they advocate, and it does not improve female performance.

A last, weak argument we shall discuss is that in secondary schools girls are offered stereotypes of mathematicians as men because they are taught mainly by men teachers. There may be more men than women mathematics teachers, but there are certainly many mixed schools with mathematics departments headed by strong women, and yet the girls do worse at 16+ and do not choose mathematics in the sixth form.

The emotional effect of bad arguments is to make one believe the reverse of what they seek to prove. That is not sensible but can be expected, and in this case it would be a great pity. Arguments can be brought which fit the facts; this does not mean necessarily they are right, but they are not subject to the same refutations that we have seen.

One argument is simple and straightforward. Mathematics is seen (probably quite wrongly) as the main measure of whether a person is intelligent or not. When girls start to go out with boys they do not wish to appear more intelligent than their boy friend. That is a heavy and unacceptable social condition indeed, but it is a fact. All desire to appear to shine at mathematics therefore disappears among the many who succumb to this pressure. What is more, the decline in performance takes place at this time; it matches the trend.

Another, less obvious argument is offered simply as a possibility. Girls are encouraged to be rule-bound and 'tutted' at when they are disobedient. Boys may be treated more roughly for disobedience, but there is a climate of approval and expectation that a boy should show 'spirit'. A possible result of this would be that girls perform as well or better when mathematics is (wrongly, it may be) treated as a set of rules. We do reach a stage in mathematics when new ideas have to break old boundaries. Breaking boundaries has been built in as acceptable for boys and not for girls.

Reasons have not been exhausted. It certainly seems unlikely that there are reasons why we cannot bring the performance of the girls to match that of the boys; the problem may well lie in the way females are treated in most societies. But beware of false arguments.

Some central uncertainties

In probability theory we are faced with seeking to predict the future, yet without absolute certainty. Gamblers through the centuries have thrived on this uncertainty. Yet in many of the unpredictable events we may feel that if we were sufficiently advanced in understanding we could know. The spin of a coin or the roll of a die is determined by the initial conditions if we know them in full enough detail, and if our physical sciences are sufficiently advanced. We may even believe that if we know completely our physical being at a particular moment and also the vicissitudes it has to face, then our moment of death is calculable from that knowledge. If we do believe such things we are taking a deterministic point of view. This view is in no way in conflict with our probabilistic attitude. Probability is a non-certain approach to problems which we can believe have certain results, which we simply are not clever enough to calculate. So we make do with statistics. In this sense, the probability does not strike at those who wish to believe in an absolute truth.

In physics the overwhelming impact of Newton through over two centuries led us firmly to believe that everything could be calculated, even if the 'three-body' problem gave Newton a headache, and we had to deal with a number of planets moving around the Sun, with many satellites. It was complexity that defeated us, not any basic uncertainty about our theories. Heisenberg struck the blow. His uncertainty principle showed that we could not determine both the position and velocity of certain elementary particles with total precision. If we found one the other was 'fuzzy'. We had a range of possibilities, not complete certainty. (This is a shattering result for those who wish to feel that at least the physical can be relied upon.) There was a probability distribution for the value we could not fix. Einstein was heard to growl 'God does not play at dice' (even though he probably didn't believe in God). This book is not about physics, but there are parallels relating to certainty that are intriguing. Perhaps this century we all became natural doubters and this invaded our science and mathematics.

There are two dominant views of mathematics (apologies to the constructivists) indicated in what we have already written. One is the Platonic, which sees mathematics as external and to be discovered. The other is the formalist, which regards mathematics as some form of game of chess, with elements with relationships between them but no dependence on the outside world. We have asserted this position in relation to testing. In particular we used it as one of the ways we extended our number concepts. Working mathematicians appear to ply their trade without too much concern for the philosophical basis. Perhaps such discussions are metamathematics rather than mathematics.

The Platonic view saw geometry as the central truth, and it was not only true, but it also matched the space of the real world. That view was destroyed, as we have seen in our chapter on geometry, by the development of non-Euclidean geometries and the fact that some seemed to fit the world better than Euclid. This did not mean that it lacked an existence and a truth of a God-given sort, in a rather more abstract way.

The formalist view will not accommodate the notion that intelligence cannot fail to develop number. We are not going to resolve these deep issues at the end of our last chapter but can state a position. Mathematics does have the objective nature, but that does not mean it depends in any way on the external world, nor does its testing reside there. That is internal. We have returned yet again to Einstein's question: 'How can it be that mathematics, a creation of the human mind independent of existence, should be so adapted to the objects of reality?'

The toppling of geometry from its perch led mathematicians to seek some new basis in number. It would be commonly held that '$2+2=4$' was one of the basic certainties. It seemed more and more of a mirage as Frege and Russell looked deeper. They sought to make sets the basis of all mathematics, and all seemed to progress well. With Whitehead, Russell wrote *Principia Mathematica*, and by page 362 had established that '$1+1=2$'. At least that suggested they were setting the foundations deep. Yet Russell then became unsure of his notion of a set, and devised statements with internal self-contradictions. He called these 'antinomies'. The original example was produced by Russell. He considered all these sets which do not include themselves; this was a notion that seemed to be both in the first category and in the second, producing a contradiction. Lest this be slightly (!) abstract consider the barber in the village who 'shaved those and only those who did not shave themselves'. At first we might pass this comment without concern, until we consider the barber himself. If he shaves himself he cannot be included among those who do not shave themselves; if he does not shave himself, then he does shave himself. Either way we are led to a contradiction. Russell conveyed this situation to Frege as he was just about to publish a major work on the foundations, which he then had to acknowledge had just given way. The hope that mathematics could be reduced to pure logic did not materialize.

A final blow to certainties was struck by Kurt Gödel in 1931 in his famous theorem. Since Euclid mathematicians had been concerned with axiomatic systems. You started with certain assumptions and built from there. For Euclid, they were 'self-evident truths' external to oneself. To a formalist they were at one's choice and had no need of external relevance, though there were internal rules that had to be obeyed. In either view it was expected that you could build a structure such that if there were a proposition it could be proved or disproved. At any stage there might be some to which we did not know the answer. Goldbach's conjecture about the even numbers is the obvious case. That does not take away from the expectation that we shall prove or disprove it at some time.

Gödel proved that in an axiomatic system such as we would need for the numbers, there would be propositions that were 'formally undecidable'. The proof is profound and perhaps one of the most shattering statements ever made in philosphy. Even in a relatively narrow area of knowledge where the starting points and the rules of the game are clear, there will be statements whose truth or falsity (within the system) cannot be established.

It is not satisfying. The hope of most people is to have something to cling on to. Faith, of course, can remain, but it would be more sense if we had a basis in reason. The answer has to be in accepting what is and not longing for what we hoped might be. There are central uncertainties, yet it is possible to live with them, and still enjoy life and mathematics.

We are now at the end of our last chapter. It cannot be hoped that we now know what mathematics is, but we must be content that we perhaps know more about it. We end with a list of quotations, unashamedly 'lifted' from the chapter headings of Morris Kline's excellent book *Mathematics in Western Culture*. Read each with care, and absorb the meaning in terms of what has been said. If, at the end, it is Huck Finn's view that most commends itself to you . . . so be it.

(1) In every department of physical science there is only so much science, properly so-called, as there is mathematics.
Kant

(2) Maths is the gate and key of the sciences. . . . Neglect of mathematics works injury to all knowledge, since he who is ignorant of it cannot view the other sciences or the things of this world. And what is worse, men who are thus ignorant are unable to perceive their own ignorance and so do not seek a remedy.
Roger Bacon

(3) Do not imagine that mathematics is hard and crabbed and repulsive to common sense. It is merely the etherealization of common sense.
Kelvin

(4) Music is the pleasure the human soul experiences from counting without being aware that it is counting.
Leibniz

(5) The science of Pure Maths, in its modern developments, may claim to be the most original creation of the human spirit.
A. N. Whitehead

(6) Geometry will show the soul towards truth and create the spirit of philosophy.
Plato

(7) Mighty is geometry; joined with art, resistless.
Euripides

(8) But where our senses fail us reason must step in.
Galileo

(9) I have never been able fully to understand why some combinations of tones are more pleasing than others, or why certain combinations not only fail to please but are even highly offensive.
Galileo

(10) For many parts of nature can neither be invented with sufficient subtlety, nor demonstrated with sufficient perspicuity nor accommodated into use with sufficient dexterity without the aid and intervention of mathematics.
Francis Bacon

(11) How can it be that mathematics, a product of human thought

independent of experience, is so admirably adapted to the objects of reality?

Einstein

(12) I had been to school . . . and could say the multiplication table up to $6\times 7=35$ and I don't reckon I could ever get any further than that if I was to live forever. I don't take no stock in mathematics, anyway.

Huck Finn

(13) Besides the mathematical arts there is no infallible knowledge, except that it be borrowed from them.

Robert Recorde

(14) Nor should it be considered rash not to be satisfied with those opinions which have become common. No one should be scorned in physical disputes for not holding to the opinions which happen to please other people best.

Galileo

(15) In order to seek truth it is necessary once in the course of our life to doubt as far as possible all things.

Descartes

(16) All the pictures which science now draws of nature and which alone seem capable of according with observational fact are mathematical pictures.

Jeans

(17) People who don't count won't count.

Anatole France

(18) Thus mathematics may be defined as the subject in which we never know what we are talking about, nor whether what we are saying is true.

Russell

Appendix

Back to basics

The squares and the odd numbers

An interesting visual way of seeing why a succession of odd numbers should add up to a square is shown below:

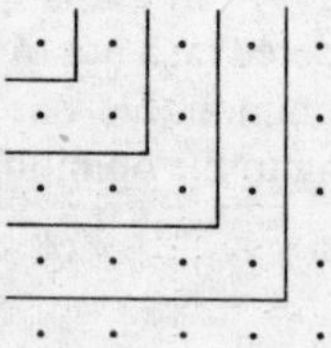

The whole array of dots is always a square, wherever we put the fence. Now look at the number of dots between successive fences. Inside the first is one dot. Between it and the next there are three, in the next gap five, and so on. This is not a 'proof', but is a fine example of 'seeing' why it is so. Seeing is one step on from pattern-spotting, and tells us something of the 'why'.
Finally, for a proof, some algebra:

$$S = 1+3+5+7 \ldots (2n-1)$$
$$S = (2n-1)+ \ldots +3+1 \text{ (just reversed)}$$
$$2S = (2n)+2n+2n \ldots +2n \text{ (adding)}$$

There are n terms, so

$$2S = (2n)n$$
$$S = n^2$$

So the answer is a square, and since we have not specified n, this is always so.

The nine times table

The amusing fact that the digits always add up to nine is an example of pattern-spotting. Again, we need to know why. The visual helps, this time with a number line.

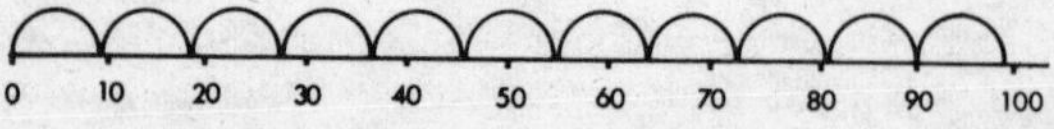

The leaps along the top are the nine times table. Each leap takes us into the next decade (ten-gap) but we fall back a unit every time because nine is one less than ten. So the tens column goes up one each time and the units go down one. Look at them:

9
18
27
36
45
54
63
72
81

Putting one column up and the other one down keeps the total constant at its starting value, nine. Investigate what happens when we get above 90, and find why the sum of the digits always divides by nine.

This fact about the nines is used in a check of multiplication called 'casting out nines'. Take this:

$$9685 \times 352 = 3{,}409{,}120$$

The digit sums of the left-hand side are 28 and 10, product 280, and of the other side 19. Taking digit sums again we find both are 10. If they did not balance, our answer would have been proved wrong. Balancing does not prove it right but it would be hard luck if we made a mistake that altered the digit sum by exactly nine.

Factorbash

The game we described appeared as an educational computer game in the US. It is easy enough for anyone to play, but deep enough to cause serious problems for a mathematician. Let us look more closely at the play with just twenty numbers. It is convenient if set up on a micro, but we shall have to make do with pen and paper.

11 12 13 14 15 16 17 18 ~~19~~ 20
~~1~~ 2 ~~3~~ 4 5 6 7 8 ~~9~~ 10

Our first choice is fixed, for whatever we choose the opponent will take the 1, if nothing else. So we take the largest prime present, 19. We shall not then be able to take another prime. Using our image of composite numbers growing from, or standing on, their prime factors, we notice that taking 9 only knocks out 3;

any other choice will wipe out two numbers. This now leaves 15 'one-legged' and we take it removing 5. We now have

11	12	13	14		16	17	18		20
	2		4		6	7	8		10

and our score stands at 19+9+15 (43). Next we tackle the powers of 2, that is 2, 4, 8, 16. Removing 16 would wipe them all out. The best strategy is to take 4, then 16. Our total moves to 63, and we now have

11	12	13	14			17	18		20
					6	7			20

We now remove the largest numbers standing on 6, 7 and 10. We add 18, 14 and 20 to our score. We have scored well over half. Our total is 115, out of the original total of 210, more than in the more casual attempt in the main text. We have used strategies based on our work on the numbers growing from primes, and trying to take numbers with only one factor left. So we know something of what we are doing, but can we positively assert that this is the maximum score for this game? That is the sort of question the mathematician asks – but cannot always answer.

Now extend the game to, say, sixty numbers. It becomes very much more intricate, yet we are still only using our multiplication tables. We think the maximum you can get is 1137, out of 1830, but a proof eludes us. Any genuine mathematician chancing to read this book will immediately wish to determine a formula for the highest score, using an array of n numbers. It should take a time, even though we are still on the basics. And then why not the problem of finding a minimum?

The auctioneers

However prices rise, if you take a percentage, then your profit remains the same in real terms. To increase a percentage in response to price rises is either foolish or knavish.

Teacher ages

Let us point to questions that arise:

(1) The number of heads is exactly the same. Does that imply an 'equal opportunity' employer?
(2) What extra information do we want, to help us interpret the numbers? The number of infant, junior, primary and secondary schools?

(3) How do we reply to the statement 'There are 50% more young women heads than men'?
(4) Why are there more young women than young men, but fewer older women than older men?

The issue really is, how many questions do the numbers raise in your mind?

The story of calculation

Gregory's formula for π

This is a good example of mathematics that is easy but 'advanced'. You need the calculus, but if you have it the process is easy.

Consider $$S = \int \frac{1}{1+x^2}\,dx$$

We substitute $x = \tan\theta$. $\dfrac{dx}{d\theta} = \sec^2\theta$

Then $$S = \int \frac{\sec^2\theta\, d\theta}{1+\tan^2\theta} = \int d\theta = \theta$$

Therefore $S = \operatorname{arc}(\tan x)$
(or 'the angle whose tangent is x')

We now return to our integral and go about it a different way.

$$S = \int \frac{1}{1+x^2}\,dx = \int (1+x^2+x^4-x^6+x^8\ldots)\,dx$$
$$= x-\frac{x^3}{3}+\frac{x^5}{5}-\frac{x^7}{7}+\frac{x^9}{9}\ldots$$

(We glide over some issues about infinite series here.)

Now put $\operatorname{arc}(\tan x)=\pi/4$, and since $\tan \pi/4=1$, then $x=1$.

Therefore $\underline{\pi/4 = [1-\tfrac{1}{3}+\tfrac{1}{5}-\tfrac{1}{7}+\tfrac{1}{9}\ldots]}$

Number

Adding fractions

It is very unlikely that you will ever, in real life, need to add fractions beyond quarters and eighths, but we shall try to understand it, just for the hell of it. We saw that the fraction 6/12 was the same as 1/2. That is unremarkable; 6 is, after all, half of twelve. With every fraction, there are whole families that equal it. Instead of 3/7 we can write 6/14, 27/63, or as many others as we choose. Suppose now that for some esoteric reason we wish to add 7/11 and 5/9. We replace each by equivalent fractions, such that they have the same denominator (bottom line). We get 63/99 and 55/99, and that is no problem. We can write 118/99 or 1 19/99, as we choose. Basically, you need only to know about equivalent fractions, and to pick those with the same denominator.

Fractpi

This is a short program which constructs fractional approximations to any number that the micro has in its memory. It does not find π, but knowing it, finds fractions near to it.

```
10 LET a = INT π
20 LET b = INT (π+1)
30 LET c = 1
40 LET d = 1
50 PRINT a;b,c;d
60 IF (a+b)/(c+d)>π THEN LET b = a+b; LET d = c+d
70 IF (a+b)/(c+d)<π THEN LET a = a+b; LET c = c+d
80 GO TO 50
```

This very soon offers us 22/7, but much more interestingly gives us 355/113, the very accurate result known to the Chinese. The curious way of going about this program hinges on the fact that if we have fractions a/c and b/d then the fraction we get by (horror of horrors) adding tops and adding bottoms of fractions lies between those two fractions. We thus narrow and narrow the gap around the number we want. This is one of the iterative processes that is made so easy by the micro. Naturally we can also approximate to numbers such as e by this method.

Geometry

Lamp-posts

The real situation with the lamp-posts is

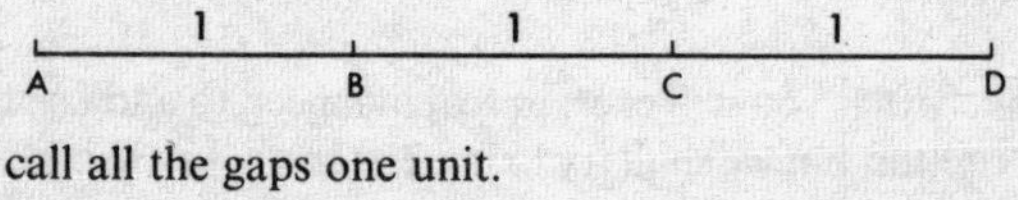

and we can call all the gaps one unit.

In projection, lengths do not stay fixed. Nor do ratios of lengths. We have to dig yet deeper, when we find that there is a ratio of ratios (called a cross-ratio) that does stay fixed. Take CA/CB and divide it by DA/DB. This number is what stays fixed. In this case it is $2/1 \div 3/2$ or $4/3$.
Now look at our projection, or perspective drawing

and we wish to calculate x. Form the two ratios.
The first is $5/2$ and the second $(x+5)/(x+2)$.
The cross-ratio is $5(x+2)/2(x+5)$.
Put this equal to $4/3$ and sort it out:

$$8(x+5) = 15(x+2)$$
$$10 = 7x$$
$$x = 1\tfrac{3}{7}$$

So we know where the next lamp-post must go.
The diagram illustrates the projection.

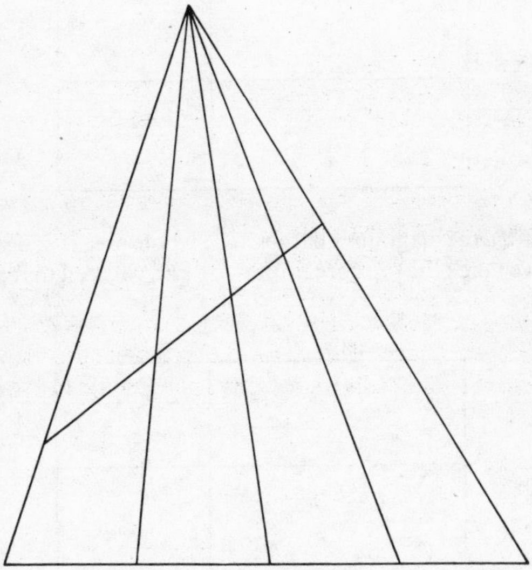

Desargues theorem

The point is that where two planes, which are not parallel, cut each other a line is formed, and all the points common to the planes lie on that line. A is in the plane of both triangles, so is B, so is C. So they must all lie on the line where the planes of the triangles cross.

Algebra

Quadratic expressions

We said, blithely, that $x^2-5x+6=(x-3)(x-2)$. If you want to handle factorizing, you need to go to an algebra textbook, but we can at least show how helpful the visual can be. Start with something simpler, say $(x+2)(x+3)$, and draw a rectangle with these sides. We have labelled the areas to establish the connection.

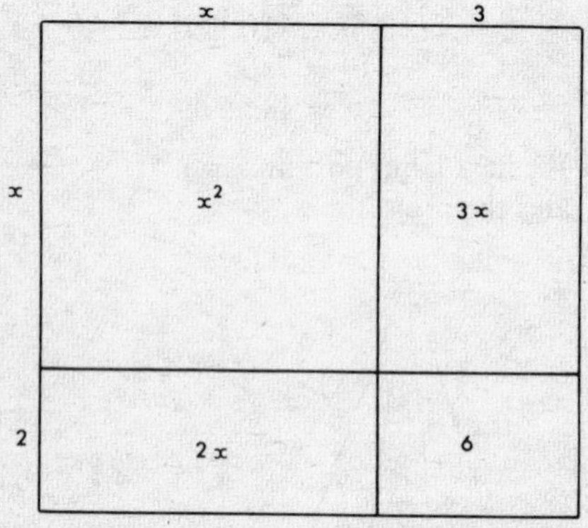

For $(x-3)(x-2)$, the diagram is harder. The whole area is x; if we take off 3x and 2x, we have taken the top corner off twice and we need to replace a 6. Try some more.

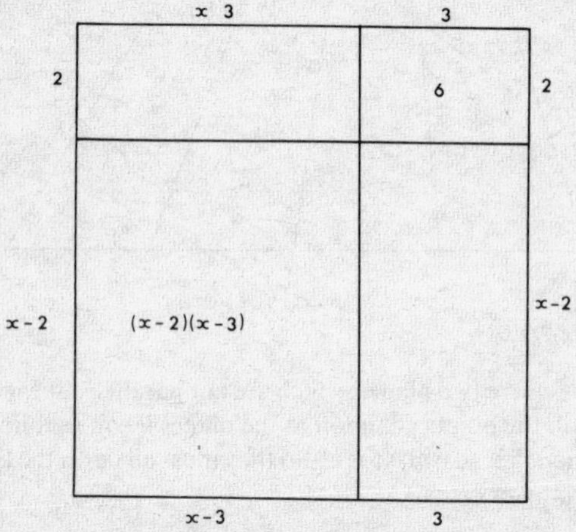

The general solution of the quadratic equations

Using general, non-specific numbers, a, b, and c, we have

$$ax^2+bx+c = 0$$

It is easier to have x by itself, so divide by a.

$$x^2+\frac{b}{a}x+\frac{c}{a} = 0$$

To get a 'balanced' equation, we need $\frac{b}{2a}$. The graph would show symmetry about that value. So we start with

$$\left(x+\frac{b}{2a}\right)^2$$

and see what this looks like. Do a picture as in our last example, if you like.

$$x^2+\frac{b}{a}x+\frac{b^2}{4a^2}$$

This is not our original equation, but by adding and subtracting we make it so, like this

$$\left(x+\frac{b}{2a}\right)^2+\left(\frac{c}{a}-\frac{b^2}{4a^2}\right) = 0$$

This can now be arranged like this, and we are at the stage when we can square root to get x by itself:

$$\left(x+\frac{b}{2a}\right)^2 = \left(\frac{b^2-4ac}{4a^2}\right)$$

Do the square root

$$x+\frac{b}{2a} = \pm\sqrt{\frac{b^2-4ac}{4a^2}}$$

and we now have got x by itself.
Rearranging we get the familiar result

$$x = \frac{-b\pm\sqrt{b^2-4ac}}{2a}$$

Similar tricks work with the cubic and fourth degree equations, but not with the quintic.

The calculus

Integration

If we want to find the area under the curve $y = x^2$ from $x = 0$ to $x = 3$, we draw rectangles inside and outside the curve. We show the effect, using rectangles one half a unit wide. The total area of these lower rectangles is (base) X (sum of height)

$$L = \tfrac{1}{2}(\tfrac{1}{4} + 1 + 2\tfrac{1}{4} + 4 + 6\tfrac{1}{4}) = 6\tfrac{7}{8}$$

The taller rectangles, and there is one more of them, give

$$U = \tfrac{1}{2}(\tfrac{1}{4} + 1 + 2\tfrac{1}{4} + 4 + 6\tfrac{1}{4} + 9) = 10\tfrac{3}{8}$$

At first sight, this does not help us much, since the limits set around the area we

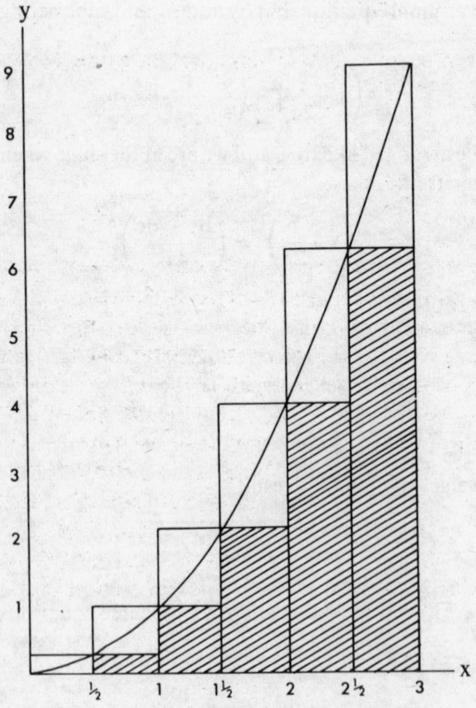

Integration: $y = x^2$

want are so wide. We do notice, though, that the expressions are the same except for the extra term on the second one.

Visionally, move the shaded steps one half unit to the left and they fit the outer rectangles, showing that the difference between the two sets of rectangles is simply that last one, height 9 and width $\frac{1}{2}$. This gives us our clue. If we made the rectangles say $\frac{1}{10}$ of a unit wide, the difference would still be one rectangle, but this time of area $9 \times \frac{1}{10}$ or 0·9. If we draw rectangles of width 0·001, and calculate, we would get the two areas only 0·009 apart. Nowadays we need not pretend to do it. We can actually do so on a micro, with a fairly simple program. We discover that the lower area is just below 9 and the upper just above, with this 0·009 gap.

If limiting processes are a worry, in that they do not seem exact, settle for being able to pin the area down very closely between limits. We can make the width of rectangles as small as we like, find the gap between the two areas immediately and find what either area is quite quickly on the micro. It is worth doing it to convince yourself.

The exponential series

When we differentiated e^x term by term, we saw that every term shifted back one.

$$e^x = 1+\frac{x}{1!}+\frac{x^2}{2!}+\frac{x^3}{3!}+\frac{x^4}{4!}+\ldots$$

$$De^x = 0+1+\frac{x}{1!}+\frac{x^2}{2!}+\frac{x^3}{3!}+\ldots$$

and we commented that the new series was the same as the old. It is slightly disturbing, for it seems to have one less term. This may not matter in a series with an infinite number of terms. Add each series, with say $x=1$, and both settle to the same result as the micro adds more and more terms.

In the exponential, however large x is, the bottom term rapidly becomes overwhelming, and the tail of the series becomes negligible. It would be convenient if there were a simple rule to tell us when the tail mattered and when it did not, but this is not so. Some series where the terms get constantly smaller do not 'converge'. The series made up of successive reciprocals 1, $\frac{1}{2}$, $\frac{1}{3}$, $\frac{1}{4}$ . . . does not converge, though its terms diminish. The other series we also mentioned, one for π, $4[1-\frac{1}{3}+\frac{1}{5}-\frac{1}{7}\ldots]$, converges, but only very slowly. We saw that after a million terms only the first few decimal places were stable.

Catastrophe theory

A number of situations in real life do not move in a continuous manner, but have sharp jerks, or 'catastrophe' changes. A smoothly moving car can undergo such a change in hitting a brick wall. The graph of its speed shows an abrupt vertical line. Suddenness does not only occur in the physical world, it is also manifest in animal behaviour. Dogs can move from cowering submission to aggressive attack very suddenly. Someone who has been learning steadily can suddenly find that they are in a state of panic.

The calculus deals with smooth changes. Only very recently did Rene Thom and Chris Zeeman devise mathematical methods to deal with at least some sudden changes, well explained in a popular manner by Zeeman in a BBC 'Horizon' programme. The mathematics is different. We cannot here go beyond this statement of the sort of situations with which it deals.

Taking a chance

Four problems

(1) We can start with any number for the birthday question. Select any one person. The next person has one chance in 365 of having the same birthday, but oddly it is the chance that she or he does not that interests us. This chance is 364 out of 365. The chance that the next person does not share a birthday with either is 363 out of 365. And so on. We then have the probability of not having a shared birthday as

$$\frac{364}{365} \times \frac{363}{365} \times \frac{362}{365} \times \frac{361}{365} \cdots$$

To get the chance that a birthday is shared we subtract the answer from 1. We need a little help from the micro. Doing the multiplication to 30 terms (the 30 people in the group) we find the product about 0·3 and the chance of there being a match about 0·7.

(2) The answer for n cards is the sum of an exponential series to n terms. As n gets large, the answer gets nearer to $(\frac{n-1}{n})^n$. For 52 cards the chance that we do not get a match is very near to $(\frac{51}{52})^{52}$.

(3) Yet again, with snake's eyes, we can calculate the chance of it not occurring, which on a single throw is $\frac{35}{36}$. So to avoid it on 36 throws is a chance of $(\frac{35}{36})^{36}$ and this again is about 0·36.

These three problems, looked at mathematically and without the surrounding story, are very much alike. The other one is different. Even when the answer is demonstrated, there is an uneasy feeling that it is not as it should be.

Assume that the three dice are labelled like this:

A: 441 441 B: 333 333 C: 522 522

There are many sets of numbers that would work, but we have chosen repeating ones to simplify the working. We need only consider three numbers on each die.

The numbers 441 win 6 times against 3 if rolled against the die with 333. (The two 4s always win, and the 1 loses every time.)

333 will beat 522, also 6 against 3. But now to our surprise, 522 beats 441, this time by 5 to 4, but a win nonetheless.

So A beats B 6–3, B beats C 6–3 and C beats A 5–4. If you choose first at any stage, the gambler can always choose the die that will beat yours.

Modern Mathematics

Part of the periodic table

5 B Boron 10·81	6 C Carbon 12·01	7 N Nitrogen 14·01	8 O Oxygen 16·00	9 F Fluorine 19·00	10 Ne Neon 20·18
13 Al Aluminium 26·98	14 Si Silicon 28·09	15 P Phosphorus 30·97	16 S Sulphur 32·06	17 Cl Chlorine 35·45	18 Ar Argon 39·95
31 Ga Gallium 69·72	32 Ge Germanium 72·59	33 As Arsenic 74·92	34 Se Selenium 78·96	35 Br Bromine 79·90	36 Kr Krypton 83·80
49 In Indium 114·8	50 Sn Tin 118·7	51 Sb Antimony 121·8	52 Te Tellurium 127·6	53 I Iodine 126·9	54 Xe Xenon 131·3
81 Tl Thallium 204·4	82 Pb Lead 207·2	83 Bi Bismuth 209·0	84 Po Polonium (209)	85 At Astatine (210)	86 Rn Radon (222)

Children and subsets

We should perhaps talk of general and specific, a rather wider idea than that of subset. The following tests, used with children of ages from 6 to 11, reveal a varied level of response, which certainly suggests different powers of organizing and ordering information. That is not to say it measures intelligence, but the ability to see what lies within which is certainly an important cognitive power.

The children are offered cards with the following words clearly printed on them

FLOWER DOG ANIMAL CHILD DAFFODIL
LEMONADE DRINK GIRL

Since the children vary so much in age and ability, it is necessary to check that they can read them. They are asked to spread them out, consider them, and then they are told: 'Sort them out in any way that seems sensible to you.' Most then do so. Some ask more questions about what is wanted; the statement is repeated and assurances given that we will be happy with whatever they do and that it is not a matter of right and wrong. Responses are as follows:

(1) Apparently random arrangements. Further questioning might reveal that, say, words beginning with 'd' are put together, and the rest put anywhere. Others may put long words together.

(2) Strategies other than the 'intended' one. The only one seen is a firm alphabetical one.

(3) Pairing of the cards. If attempted with these cards, the pairing always seems to be the 'normal' one. The only exception is with sophisticated adults determined to show how clever their thoughts are!

(4) Arranging in rows or columns. This structuring seems an extra element, other than the general and specific. Mostly they are side by side pairing like this:

DOG	ANIMAL
FLOWER	DAFFODIL
CHILD	GIRL
LEMONADE	DRINK

Sometimes they are arranged 4×2 rather than 2×4.

(5) If asked at this stage whether they are 'happy', children switch cards in columns. Some are simply trying to please and look at you to see facial reaction. Others appear to have a principle – but not a clear one.

(6) Arranging in columns, with general all in one and specific in another. Lest this be by chance, it is then necessary to ask what they have done. The clarity of explanation can vary greatly.

This simple testing shows something of a developing cognitive facility. It can be extended, with further cards structured in the same way, but dealing with number and shape. There is then the added issue about whether the general and specific always appear the same way round, and the clear observation that the patterning and process is the same in every case, though the material is different.

Valid and invalid argument

The most common form of invalid argument arises from not knowing clearly what lies within what, as with our footballing fools, and it is easy for any of us to fall into error. The use of enclosures here is very helpful, particularly for those to whom the visual appeals. We now offer a series of invalid arguments, and suggest that you assure yourself that they are invalid and draw diagrams to illustrate it.

(1) It is raining again today. It always rains on Monday. So it must be Monday again.
(2) Good cars are expensive. I paid a lot for my car, so it must be a good one.
(3) Some clever people have degrees. John has a degree so he must be clever.

The interesting point about these arguments is the distinction between the truth value of individual statements and the process that leads from the first two to the third. In asking whether an argument is valid we do not need to discuss if the statements are true. Consider the following syllogism:

'Only boys are clever. Jane is a girl. Therefore Jane is not clever.'

The logic of the process is impeccable. It may happen that you do not agree with the first statement, but that is a matter outside our brief in discussing process.

How is your logic today?

Some of these are just for fun, but let us offer some answers.

(1) Contrary to a developing usage, 'verbal' covers both the written and spoken word (even if the police do not believe so), so the answer is III.
(2) Something of a catch. There is an even prime, 2, so we must pick II.
(3) People can play both, but do not have to . . . II.
(4) Sixth powers of numbers, such as 64 which is 2^6, or 243 which is 3^6, are both squares and cubes . . . II.
(5) A square is a special sort of rectangle . . . IV.
(6) Well then? Is price one aspect of value? Up to you.

(7) Reluctantly, let us admit that calculation is an aspect of mathematics . . . III.
(8) It depends on how you view Y.
(9) There are a lot of points on a line. A mathematician might respond that there are a lot of lines through a point. Still, let us go for III.
(10) Double bluff. As easy as it looks . . . IV.
(11) The readership of *The Times* is wider than the top people in this statement, but did the advertisers just hope that you might be tempted to buy in the belief that that might make you top?
(12) Bulmers is a make of cider. By suggesting it is also the other way round, they are saying that both are identical. That would be a proper conclusion as far as process is concerned, but you are not obliged to believe the first statement.

Waltzing ducks

The point here is that we are not tempted to discuss the truth value of the statements. We can therefore consider process by itself.

(1) Ducks do not waltz.
(2) No officer ever refuses to waltz.
(3) My poultry are all ducks.

The ducks and waltzers are kept firmly apart in the first diagram, expressing the first statement. The next statement is more tricky. If they never decline, it must be that they all waltz, so our next diagram effectively says that all officers waltz. Of course others may waltz as well.

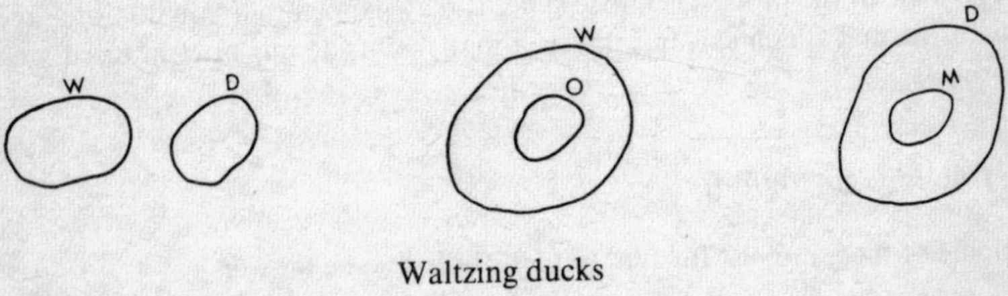

Waltzing ducks

The third statement also has a little difficulty. If we regard the sense (which we should not) we are worried that ducks are in fact a subset of poultry. But not 'my ducks'. Look at the diagram. We now draw the compelling conclusion that 'My poultry are not officers'.

The conclusion does not enlarge our knowledge of the world, but it is a proper conclusion. It can certainly be reached without diagrams, yet it is surprising how often those who depend just on language reach wrong or, more commonly, incomplete conclusions. It is true, for instance, to deduce that 'Ducks are not officers' yet it draws on only two of the statements, and is hence incomplete.

If you feel that language should suffice, attempt to draw a single conclusion from these:

(1) Every idea of mine that cannot be expressed as a syllogism is really ridiculous;
(2) None of my ideas about Bath-buns are worth writing down;
(3) No idea of mine that fails to come true can be expressed as a syllogism;
(4) I never have any ridiculous idea that I do not at once refer to my solicitor;
(5) My dreams are all about Bath-buns;
(6) I never refer any idea of mine to my solicitor, unless it is worth writing down.

Fortunately it is only those who enter mathematics or the law who have to bother with this sort of thing.

Mathematics in action

Secondary school placement

More complicated allocation systems try to take into account various criteria. The following description is broadly based on that used by the Inner London Education Authority.

The first criterion is parental choice, an admirable policy, even if it sometimes stimulates expectations that cannot be met. The next is 'balance of ability', on which we shall need to elaborate. The third is nearness to school, which corresponds roughly to the catchment idea, but relegates it to third place.

In the first stage, therefore, after discussions between the parents, their children and the primary school head, the parents state their preference. Not unnaturally, this results in some schools having more who choose them than they can accommodate, though a surprisingly high proportion of people do get their first choice. This is what constraints in a system are about, however. Not everyone can always have what they want. We cannot build extra rooms onto a school which one year attracts more than it can hold.

The next criterion is balance of ability. If you run a comprehensive system, you cannot, for example, allow a disproportionate number of the most able to attend a certain school. If a school, justly or not, has gained a reputation for its more academic work, parents of more able children would, not unreasonably, seek to choose it. This would further enhance its results, and the process would continue. A 'banding' policy overcomes this. If in primary we allocate 25% to band 1, 50% to band 2, and 25% to band 3, and then ask that secondary schools admit in these proportions, then they become balanced in their clientele. So a school which takes in 240 children a year is asked to take 60 in band 1, 120 in band 2, and 60 in band 3.

A patently fair policy does not always appear so to the customers. Since the decision on who is turned away (because a certain band is full in that school) hinges on distance from school, though that is the last criterion applied, there are apparent anomalies. A school thought to be 'academic' may attract too many band 1 applicants. Some are then turned away, while band 3 children from further off are admitted. An obvious consequence, not easily understood by all.

In Inner London, upwards of 20,000 transfer every year, and it is clear that not all the criteria can be met. The more mathematical task is to minimize the breaking of rules. To illustrate this we invent a small problem based on the real one. Our table shows the situation of three children and three schools. We are told of the parental choice, the nearness to school, and if there is room in the right band. We now add an artificial constraint that each should go to a different school. The problem, though much reduced in size, retains many features of the real one. It does not take long, by trial and error, to find that we cannot fit them in under the conditions laid down. So we must break some rules.

	Meadway			Dell			Arbour		
	Parental choice	Nearest	Ability band	Parental choice	Nearness	Ability band	Parental choice	Nearness	Ability band
Ann	1	2	√	2	1	×	3	3	×
Barry	1	3	√	3	1	√	2	2	×
Charles	2	1	√	1	2	×	3	3	√

Put the Children in School

Our next diagram introduces a system of penalty points. Putting a child in its second choice costs one point; in its third choice two points. We use the

	Meadway			Dell			Arbour		
	Parental	Nearest	Ability band	Parental	Nearness	Ability band	Parental	Nearness	Ability band
Ann	0	1	0						
Barry				2	0	0			
Charles							2	1	0

Total 6 penalty points

Our placement of children, Ann in Meadway, Barry in Dell and Charles in Arbour costs 6 pts. Can you find a placement that costs less?

same principle for distance. And if we dare (*pace* the bureaucracy) place a child where there is no room in the band we again lose a point. Now the job is to find the placement of the three children that scores the least penalty points.

The virtue of an approach like this is that any decision at all can be made about penalties. For instance, a very much higher number of points may accrue for giving only the third choice of school. Also, with proper computer backing, the general problem can be tackled along very similar lines.

Playing a game like this brings home the constraints that are normally present in real issues, and shows us the way that mathematical ideas can help in community matters.

Magistrates courts rotas

First let us describe the essential features of the way a magistrates court works, using a typical court as an example.

Let us take a court with 75 magistrates (or justices, or JPs, as they are variously called). Let us suppose all of them sit in the main courts, which deal with motoring offences, theft, some cases of assault, indecent exposure, fiddling the railway, and so forth. Even a mass murderer has first to appear before a magistrates court, before he goes 'up the road'.

Normally a 'bench' consists of three justices, one of whom is designated chairman. He or she controls procedure in court, but all have an equal say in conviction and sentence. Practice about chairing varies. In some courts it seems to be almost by lot, but in most there are clear policies as to who does and who does not chair. In the court we are taking as an example, a rota committee decides who chairs and how often, and this largely hinges on seniority. No one chairs until they have been on the bench for several years, and then infrequently. The most senior tend to chair nearly every time they attend, though sometimes they sit under a new chairman to give help. Our rota has to take account of all these matters.

In most parts of the country the juvenile panel is elected from the main bench, with the intent of getting the younger people, with experience of adolescents. This is not a subsidiary job; if anything it involves more not less responsibility. In our court there are 27 members. Among themselves they elect a chairman and a number (decided by them) of deputy chairmen. Only these people are allowed to chair in juvenile. They may or may not be people who chair in the main court. We can see the problems building up for those who make up the rota. The pattern for the Domestic Panel is exactly the same, but the decisions they make on the number of deputies may be different. Finally there is a bench which deals with betting, gaming and liquor licensing; it meets only five or six times a year, but is yet another sub-set of the main bench.

Accustomed as we now are to Venn diagrams, we draw this situation. People may sit in any or all of the 'sub benches'. To sit on all may mean that attendances on the main bench become too infrequent, but there is no legal bar. It would now be possible to cut up the sets according to whether people could chair or not; the diagram would look rather more complicated, particularly since some can chair in one court but not another. Grasping the various categories is essential to tackling our problem.

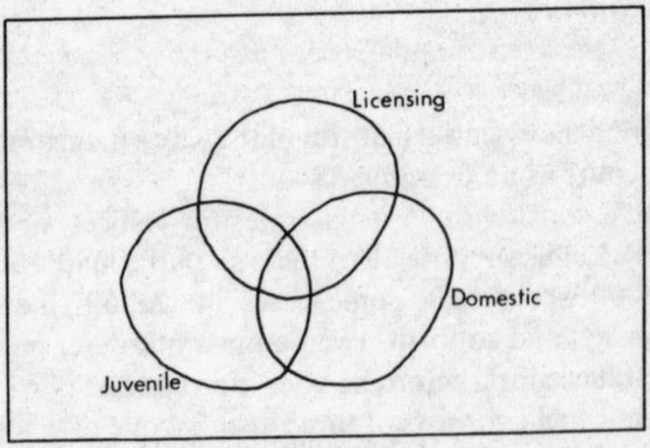

Chairing courts

The situation of chairs in main court is worse than in the panels, since we need to programme those who chair for a different number of times, depending on their seniority. Already the number of constraints seems considerable, yet we now add a further series of conditions.

(1) All benches in juvenile and domestic must have a member of each sex on them.
(2) It is desirable that the same condition should obtain on the main bench.
(3) All justices should be programmed to sit in an even pattern throughout the year, and never twice in a single week.
(4) It is necessary that everyone should meet everyone else. There should not be regular benches of the same people.
(5) Every week a justice or two will be needed to sit at Crown Court. No one should be chosen who is already sitting that week.
(6) Everyone should sit more or less the same number of times in the four court rooms.
(7) Afternoon courts are a special feature, and we need to take into account those who wish to sit in them.

We thus have a large number of constraints. Some courts may add to the difficulty by allowing people to exercise individual preferences. This is not the situation in the court we are looking at; there, a year's rota is prepared well in advance, and people are expected to swap to meet their special needs. This can of course alter balances and regularity, but does not matter since it is at their choice. It is necessary to change with someone in the appropriate subset, however. A non-chairing woman in juvenile changes with another such.

The interest in devising such a rota is twofold. We want to meet all the requirements, and we want to establish, not a rota for next year, but a general method that may always be used. Essentially we need an algorithm, and if we get a suitable one, we shall be able to put it on a computer. The technique is to tackle one issue at a time, and it requires intuitions to decide which to tackle when. Intuitions are not mystical. They stand on previous experience and mathematical insight.

The first problem is to establish patterns that ensure both evenness of attendance and constant mixing. These two criteria operate against one another. The simplest way to get evenness is to write everyone in a line and string them in. However, it would mean that certain benches constantly recurred.

We use the prime numbers in a new way. Take a prime number of numbers! For instance

1 2 3 4 5 6 7

Where we have taken seven numbers

Vary the pattern by picking every second one, thus

2 4 6 1 3 5 7

Now pick every third, every fourth, and so on. We get six strings of numbers. Selecting them in a careful order, writing them in one long list, we have a string that is mixed, but reasonably evenly spaced. We can do this with any string of numbers, but the primes are the easiest to use. We now establish these 'circulation patterns', for a number of primes, probably up to 19. Once established they form a bank on which we shall draw.

We now allocate numbers to magistrates. These are not simply in order, though, but code the various sub-groups. For instance, we may arrange that all juvenile magistrates have numbers from 10 to 30. We also allocate odd numbers to men, and even ones to women (or vice versa).

The next stage is to make a rota for the domestic and juvenile panels. We need to deal with these four sub-groups:

(1) Juvenile chairmen.
(2) Other members of the juvenile panel.
(3) Domestic chairmen.
(4) Other members of the domestic panel.

Suppose there are 16 non-chairing juvenile justices. Attach their numbers to the circulation pattern for 17, leaving out 17 itself, and we have all the magistrates in this group in a long, repeating string, satisfying our conditions. In a table for the full year, we simply enter the string in, repeating it if necessary.

This process is done for all groups, no attention being paid to clashes in a week arising from those who sit on both panels. However, those who do, have their numbers ringed, to make them easy to pick out. Nor do we bother about the sex balance. This is characteristic of seeking an algorithm; we need to do one thing at a time, and if we try to fit all the conditions at one go we shall get in a muddle. Now, however, with all the allocations made, we do sort out clashes; it was an intuition that this would not be a very extensive job. In fact a dozen or two moves sort this out. Checking that we have a sex balance simply means looking at odd and even numbers. Again, relatively few changes are needed.

The licensing courts are so few that fitting them in at this stage causes little difficulty. The final trick is a nice one. Instead of allocating the main courts we allocate the free time, and allocate that evenly. This leaves the remaining magistrates free for those courts, and there is much flexibility in how we then attach to individual courts. As with the panels, the allocation is a block one for a week, but with the assurance of evenness, mix, sex balance and no clashes.

There is more detail, but we hope the plan is clear. In arriving at our algorithm we used the primes, some sets, odd and even, and took a complement at one stage (the free time) instead of the set itself.

Wine bibbing

A relatively short experiment will enable us to find an answer, probably by trial and error. But a geometric approach with points and lines offers a routine which will always work. In our diagram we use 'isometric' paper which has a grid of equilateral triangles over it. We draw axes at 60° to one another with lengths 5 and 3 along them. We can plot points on this paper as easily as on the more usual square grid; for instance the point C has coordinates (3,3). We shall interpret this as meaning that the 3 and 5 gallon jugs both have 3 gallons in them.

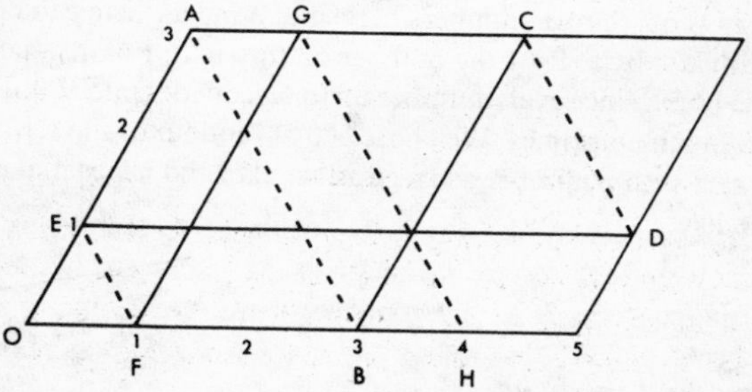

We start at (0,0), and using the 8 gallon jar as reservoir we fill either one of the others; here we have filled the 3 gallon one. Now from A we bounce round this odd-shaped billiard table on the path A to H. On this path the dotted lines are where we pour from the 3 to the 5, and the continuous ones where we empty into the reservoir or fill up from it. The points we hit on the route give the possible amounts in the two jugs, and when we reach H we have 4 gallons in the jug and the problem is solved.

The reader may care to try filling the 5 gallon one first, and seeing what path we then get, or try different jug sizes.

Algorithms, problems and investigations

A prime problem

Consider the number 600! The exclamation mark is not part of the punctuation; it has a mathematical meaning. For example, 5! means $5 \times 4 \times 3 \times 2 \times 1$ or 120. By the time we get to 8! we are already in tens of thousands. A newspaper headline once read 'Bradman out for 14!' If the exclamation mark had our present meaning, the score would indeed be remarkable, even for Bradman. For 600! we therefore have to take $600 \times 599 \times 598 \times 597 \ldots 3 \times 2 \times 1$, which is rather large. Now consider the string of consecutive numbers:

$$\begin{array}{c} 600! + 2 \\ 600! + 3 \\ 600! + 4 \\ \\ 600! + 600 \end{array}$$

There are 599 consecutive numbers here, and none of them can be prime, for every one divides by at least the number on the right. For instance, $600! + 59$ must divide by 59, since every number up to 600 divides into 600! This gives us a gap of 599 with no primes. We chose 600 at random to start. It is clear that we could start with a number as large as we like, and hence make the gap as large as we like.

The hunter and the bear

After the satisfaction of finding the North Pole solution it is a bit of a letdown to be told there are more answers. Despite one's natural suspicions, there are more answers, genuinely satisfying the conditions. Move a short way out from the South Pole to where the line of latitude is exactly one mile round. Now find the line of latitude that lies one mile north of that. From any point on that circle, a mile south takes us on to the smaller circle. Travelling a mile along that brings us back to where we started on it. Then for the last mile, we retrace our steps up to our starting place. Again, we sit back, well satisfied, only to be asked 'And where else?' It is beginning to look like a Russian doll.

We now take a whole set of smaller circles, with circumferences $\frac{1}{2}$, $\frac{1}{3}$, $\frac{1}{4}$, . . . and so on. The solutions are the circles lying a mile north of each of these. Our problem started with a single point answer, extended to an infinite number of points on a circle, and then to an infinite number of circles. Always make sure you have finished.

Cutting the cube

There are 27 small cubes. The central one needs all six faces cut, and they must be cut separately. So six is the minimum.

Intersecting lines

The various special cases of the three lines are shown below:

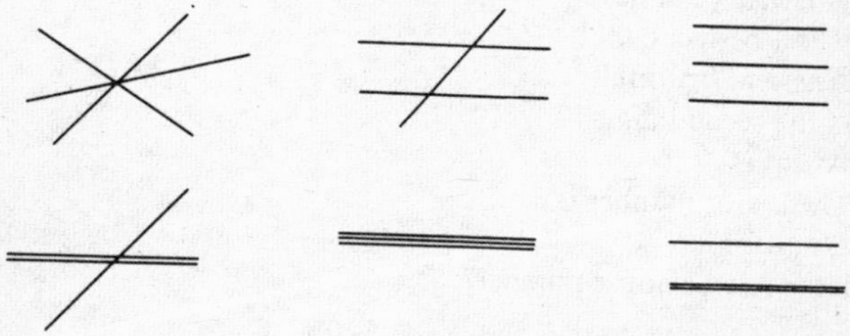

The special cases with four lines are more complicated, but based on the same principles of concurrency, parallelism and coincidence of lines. The four-line configuration is unique, leaving aside special cases. Take the three lines and add the fourth. Effectively there are only two cases, where the line cuts inside or outside the triangle. Both give the same configuration.

Some facts are reasonably easy to sort out. The number of open regions is twice the number of lines. As we add a line it cuts an open region on each side, adding one to each, which establishes it. Closed regions are slightly more difficult, but starting from 3 lines we get 1 closed region, and the numbers 1, 3, 6, 10, 15 . . . which are known as the triangular numbers (think of the reds in the snooker triangle).

Some more problems

Books of games and puzzles in mathematics abound, but we offer you a small and amusing collection, roughly arranged in order of difficulty, though this is very subjective.

(1) A large conference is held, and as people arrive, they shake hands with those they know. Show that the number of people who shake hands an odd number of times is even.

(2) A house has a number of rooms, all the doors either connecting with another room or with the outside. It is found that every room has an even number of doors. Show that the house has an even number of doors to the outside.

(3) Alan lives at a house whose number is between 10 and 1300. Bert knows this fact and asks him some questions:

B: Is it above 500?
A: Answers (and lies).
B: Is it a perfect square?
A: Answers (and lies).
B: Is it a perfect cube?
A: Answers (truthfully).
B: Is the second digit 1?
A: Answers.
B: Then your number is . . . .
A: No, it isn't.
What is Alan's house number?

(4) A man bought four articles at the supermarket and entered them correctly on his pocket calculator (say 1·53 for £1.53). Instead of pressing the addition, he pressed the multiplication sign each time. Getting the answer 7·11, he went to the checkout, where the girl entered the amounts correctly, added them properly, and accepted his £7.11 as correct. What were the prices of the four articles?

(5) You have 12 coins, one of which you know to be dud and to weigh either more or less than it should. You have a simple balance (no readings) and using the coins as weights, you have to find which coin is dud, and if it is light or heavy.

(6) In the triangle AB = AC. Using Euclidean geometry, find the marked angle.

(7) Find all the triangles whose sides are whole numbers, and whose area numerically equals the perimeter.

(8) Can a cubical box be exactly filled with a finite number of other cubical boxes, all different?

Some investigations

It takes practice and confidence to embark on investigations. The initial reaction is 'What am I to do?' It is up to you. Devise questions in the area suggested, and then solve them. Once you get the idea, it is a relief not to be told what you have to do. In the following examples, some questions are raised to start you off, but ignore them if you wish.

(1) Examine planes intersecting in 3-D space. As before, look at the regions created, possible configurations, the number of lines and points formed, how many special cases there are, and so on.

(2) Investigate points on the surface of the Earth. For instance, how many sizes of 'equilateral triangle' can you get? Remember to use great circles for straight lines. With more points, can they be spaced with all the distances between them being simple multiples of one another? Plan some shortest distances taking in several cities in a continuous route. Discuss what 'going round the world' means.

(3) In the Syracuse conjecture we started with any number, and took two particular rules. If even, divide by two, if odd, multiply by three and add one. We always return to one. Try other rules, and examine the strings of numbers.

(4) Preferably using a micro, examine the gaps between primes. Find the first decade with no primes (possible without electrical help). When is the first time we get a gap of 50? Look at the distribution of prime 'twins', which are pairs of primes separated only by 2.

(5) Investigate visiting points in a plane. Start with four, and see how many routes there are to visit all four, starting at any one you like. Which is the shortest? Is there a rule? Remember the answer could be no.

(6) Given a square of side, say 17, examine ways of cutting it up into small squares all different. If this is not possible, can just one rectangle be left?

Mathematics: its nature and purpose

Popper's three worlds

Sir Karl Popper is one of the greatest influences and one of the most controversial of modern philosophers. He seeks to tell us what science is about, and in so doing uses a classification of 'all there is' that can stand on its own and offer illumination as to the nature of mathematics. In our discussion on sets we said that we could classify how we chose; that there is not a truth issue involved. Nonetheless some definitions seem to have more mileage in them than others, and to lead more easily to further thinking.

An important issue in philosophy is the mind-body problem, and arguments have raged between monists and dualists for a very long time. Popper regards the physical and mental as conceptually distinct yet interactive. He defines World 1 as the physical world, not only of concrete objects but of forces and fields. He does not seek to question whether this 'really' exists. World 2 is our inner world, a world of mental states of consciousness and psychological dispositions, and unconscious states. This classification has nothing new about it, though the position of 'interactionism' he adopts is a significant development. There is no assumption that one could exist without the other.

It is in World 3 that he introduces a new idea, useful to us here. We shall use his own definition from *The Self and its Brain*: 'by World 3 I mean the products of the human mind, such as stories, explanatory myths, tools, scientific theories (whether true or false), scientific problems, social institutions and works of art'. This particular definition is new and exciting. In the not impossible event of our wiping ourselves out, an alien intelligence visiting this world in the future will know us. By our works shall we be known. This idea takes a time to get used to. It is easy to define, but to become accustomed and attuned to an idea we need to work with it. Popper illustrates it with a book. The physical object is in World 1. Now take a second copy of the same book. It is a separate object in World 1, but its content is the same as the first book, and so there is only one World 3 object. It does not matter that different people will reach different interpretations of the meaning. That is the result of interaction between their different World 2s and a single World 3 object. The two copies of the book certainly make the same statement whether we agree what that statement is or not.

It may help to stabilize the idea of the three worlds if we apply it to some of the things we have been talking about. The body of mathematics is in World 3. It is a construct of the human mind, externalized and available to all once uttered (published in some way). The 'real' world of which we have been speaking certainly contains all of World 1, but there are aspects of it that lie in

World 3. The 36 bus is in World 1, but the arrangement whereby it travels from your house to your place of work is a human one and lies in World 3. Mathematics may be used to act directly on the real world, or it may act indirectly by first acting on other World 3 material (the other subject disciplines), which then acts on, or at least interprets, the real world. When we regard mathematics as useful for personal development, we hope that the interaction of our own World 2 with this World 3 material will benefit our World 2.

This is the first of two frameworks we shall use to further the discussion of the nature of mathematics.

In his *Intelligence, Learning and Action* (1979) Richard Skemp discusses reality testing. He says:

> 'The accuracy and completeness of our conceptual structures are tested in three ways:
> (a) by experiment
> (b) socially, and particularly by discussion
> (c) by internal consistency.'

While he (despite this author's urging) does not accept the idea of World 3, we see a considerable agreement between the positions. Skemp is asking that we test our ideas against World 1, against other World 2s, and internally within its own part of World 3.

Armed with these two conceptual frameworks we shall now attempt to say something of what mathematics is, though a definitive answer is hardly to be expected. There are different ways of looking at things, standpoints that give new insights, and that is what we aim to offer. We shall treat as significant what a particular study is about (Skemp uses the term 'operand'), and the manner in which it is tested. These two criteria lead us to classify subject areas in ways not general in our educational institutions.

The scientists concern themselves with World 1, and for those of them whose position is firmly materialistic that is all there is anyway. Not only is science the study of World 1 but, critically important, that is where the testing must lie. Popper asserts the importance of 'falsifiability' in a scientific theory. Essentially a scientific theory must predict what will happen, and afford the scientist opportunities to conduct experiments which may disprove the theory. No amount of experimentation confirms a theory; a single experiment may disprove it. Viewing the negative side in this way does not appeal to some, largely on grounds of temperament, yet it is the stronger position. In mathematics too, which we shall not classify as a science, the single disproof is a powerful weapon. Were we to find a single case of an even number which could not be expressed as the sum of two primes, Goldbach's conjecture would be finally refuted.

When a theory survives certain tests then it does make us more willing to accept it emotionally, even if rationally we are no nearer certainty. Einstein made certain predictions concerning the way the planet Mercury moved. These predictions were related to the bending of light rays near a massive object such as the sun and involved firm numerical answers to certain experiments. When these were done he was found to be more accurate than Newton. This did not prove his theory, but the move towards his thinking gained great momentum as a result.

A consequence of the principle of falsifiability is that we can never prove anything in scientific theory. This is in no way disabling, for despite the knowledge that the theories with which we work will in time be superseded by others, the predictive power of science is impressive in the extreme. Even if we now believe Newton's resolution of the physical world to be wrongly based, it can still be used with total confidence to give very accurate predictions in many areas. The characteristic of the sciences, then, is that they are concerned with World 1 and are tested there. We start another series of diagrams illustrating our classification.

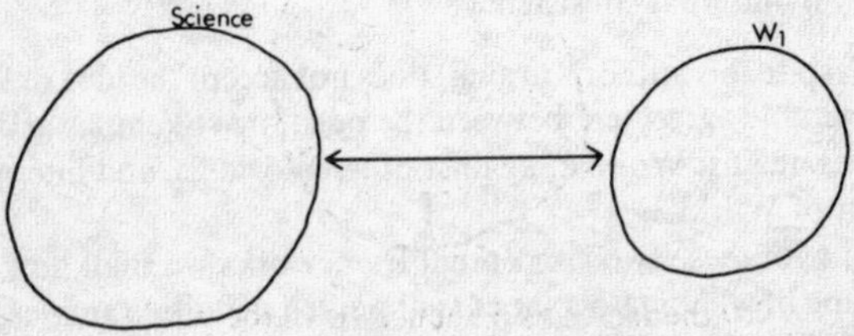

Science and World 1

More recent studies, such as psychology, are often treated as if they should be developed in the same way as, say, physics. It is for this reason that behaviourism gained so much ground and the stimulus-response model sought to reduce people to billiard balls. Psychology is not about World 1, but World 2. If it is about something different from science, we shall not call it a science. Testing is presumably also in World 2, even if we deduce what happens in another's mind by watching their behaviour.

We have been at some pains to indicate the power of mathematics in the real world, but unlike the sciences, that is not its area of study, not what it describes. Despite its range of applications, it cannot properly be said to be about anything (as the quotation from Russell suggests) except perhaps itself. It is in Einstein's phrase '. . . a creation of the human mind independent of experience . . .'. He points to its detachment from that world despite its power as a language of explanation of that world.

The second question we have been posing about various studies is 'Where is it tested?' Here again mathematics has a deeply different nature from

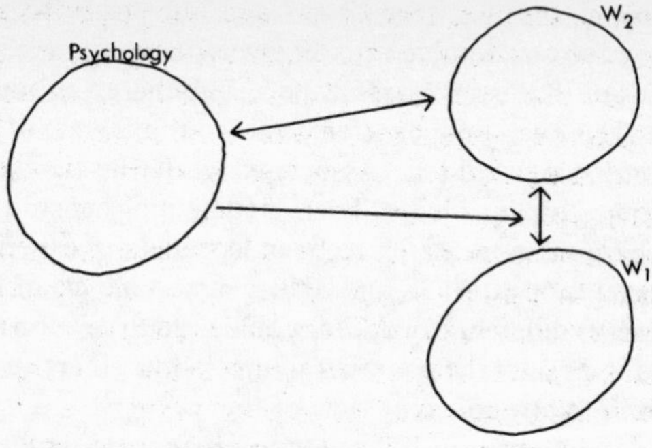

Psychology, World 1 and World 2

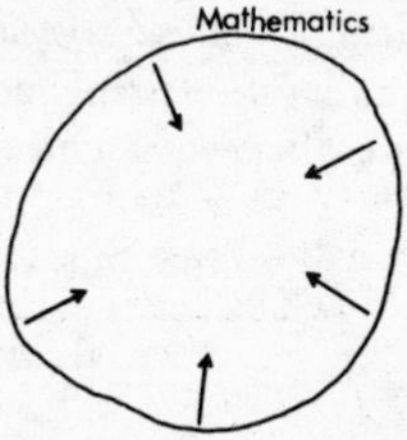

Mathematics as detached from the world

science. Tests of mathematics do not lie in World 1, they lie within mathematics. The test for validity is internal consistency (Skemp's (c)), and is therefore in World 3 and not World 1. The subject is almost self-referent.

Different starting points might yield different views as to which subject is most closely associated with another. It is a matter of our own judgement what criteria we use, yet surely the answers to the two simple questions we have been asking must have relevance. If the answers are distinct for two subjects, then clearly the closeness of their relationship is in doubt. So the fact that an advanced work on theoretical physics seems to contain nothing but mathematics does not mean that there is any deep connection. It merely means that mathematics is the language in which we write physics. A history text is not a work on English because it happens to be written in that language.

Mathematics illustrates some very interesting points about World 3. If we take the somewhat risky step of saying that we invented the counting numbers, and that as utterances of the human race they are part of World 3, how does it come about that there are things about them that we did not put there and that we have discovered? These may be simple matters, such as the fact that adding a string of successive odd numbers, starting at one, always gives a perfect square, or problems that we have enunciated but not solved, such as the question whether two consecutive squares always have a prime number lying between them. None of this sort of issue did we put there; the numbers appear to have a life of their own. Popper refers to this curious state of affairs as the 'partial autonomy' of World 3. Associated with this is the claim that mathematics is objective knowledge. There is something here too approaching a paradox. If we claim mathematics to be independent of experience, and its tests to be internal, not external, and at the same time claim that an alien intelligence, however different from ours, would need to develop number, then we are in a bind. Not only the counting numbers but others such as π and ε seem to be built in to any coherent view of things.

A science fiction story by Isaac Asimov makes the point well. Radio messages comprising sounds and gaps are received from outer space. They recur with a period of 1681 so naturally the mathematicians set them out in a square 41×41 (everyone knows that $1681 = 41^2$). The picture tells, in numbers, where the planet of origin lies in its planetary system, what its main chemical constituents are, and various other interesting facts. It suggests that the only language we shall be able to use when we first meet other intelligent life will be mathematics.

We may not have resolved anything, but we hope it was fun.

Understanding

The answers given to this series of 'explanations' may reflect as much the way a person has been taught as their understanding. The first possible reply is the parrot-like rule that some of us may remember. It is actually wrong. If you add 0 to anything you leave it unaltered. We think of this, however, as 'instrumental'. The second implies an understanding of the sort we seek, and contains a valid reason for what we do. We call it 'relational'. The next explains nothing and the last is nonsense. Think about these different explanations, and consider how they reflect understanding.

Some suggestions for further reading

There are many books full of mathematics, yet impenetrable to anyone save a specialist. In searching a library, beware of those books with 'Elementary' in their titles. You usually need a maths degree even to make a start on them.

Perhaps the starting point for books 'about' rather than 'of' mathematics is Lancelot Hogben's *Mathematics for the Million*, the first book to make a real impact on the public. Many a young person must have had his or her interest caught by this book, and been influenced by it in future educational choices. Another classic is W. W. Sawyer's *Mathematician's Delight* (Penguin). Warwick Sawyer is an engaging personality and an elegant writer. All his books are well worth reading.

No one should miss Morris Kline's *Mathematics in Western Culture*, with its strong historical theme. Rather more technical, and needing some background in mathematics, is Courant and Robbins's *What is Mathematics?* Lastly in this short list of books about mathematics is Norman Gowar's *Invitation to Mathematics*. The author is now Professor of Mathematical Education at the Open University, and was my co-presenter in the BBC TV series 'Maths. Help'.

For those who want to know what it is like to be a working mathematician the seminal work is G. H. Hardy's *A Mathematician's Apology*, which has held the field for a long time, but is now supplemented by *The Mathematical Experience* by Davis and Hersh (Harvester).

For the discussions on graph theory in this present book, I am much indebted to my old friend and former pupil Adrian Bondy. For further reading in this new area of mathematics we have *Graph Theory with Applications* (Bondy and Murty: American Elsevier). Though this is a technical book, throughout it deals with practical problems, and has a solid grasp on reality.

Those whose interests lie on the psychological or educational side will probably have read *The Psychology of Learning Mathematics* (R. R. Skemp, Penguin), but if you have not, do so! Richard Skemp is Professor at Warwick, and an old friend and co-worker of mine, who has greatly influenced my thinking. I hope you might also read my study of how people become disaffected with mathematics *Do You Panic About Mathematics?* (H.E.B.)

If you are interested in puzzles and problems, then read any of Martin Gardner's excellent books published by Penguin.

It would be possible to suggest many more – names and titles flood in – but it is important not to be overwhelmed. It is my belief that any intelligent adult should be knowledgeable about mathematics, as they should be about other aspects of our culture. I further believe that those working in mathematics in any way need to see it in context. It is all too easy in our present educational system to pass very advanced examinations in mathematics and know nothing of its philosophical importance, and little of its impact on society.

Index